Identification of Continuous-Time Systems

ENGINEERING SYSTEMS AND SUSTAINABILITY SERIES

Series Editor: Ganti Prasada Rao

Co-Editors: D. Subbaram Naidu, Hugues Garnier, and Zidong Wang

Published Titles

Identification of Continuous-Time Systems: Linear and Robust Parameter Estimation
Allamaraju Subrahmanyam and Ganti Prasada Rao

Nonlinear Stochastic Control and Filtering with Engineering-Oriented Complexities
Guoliang Wei, Zidong Wang, and Wei Qian

Multi-Stage Flash Desalination: Modeling, Simulation, and Adaptive Control
Abraha Woldai

Identification of Continuous-Time Systems

Linear and Robust Parameter Estimation

Allamaraju Subrahmanyam
Ganti Prasada Rao

CRC Press
Taylor & Francis Group
Boca Raton London New York

CRC Press is an imprint of the
Taylor & Francis Group, an **informa** business

CRC Press
Taylor & Francis Group
6000 Broken Sound Parkway NW, Suite 300
Boca Raton, FL 33487-2742

© 2020 by Taylor & Francis Group, LLC
CRC Press is an imprint of Taylor & Francis Group, an Informa business

No claim to original U.S. Government works

Printed on acid-free paper

International Standard Book Number-13: 978-0-367-37143-2 (Hardback)

This book contains information obtained from authentic and highly regarded sources. Reasonable efforts have been made to publish reliable data and information, but the author and publisher cannot assume responsibility for the validity of all materials or the consequences of their use. The authors and publishers have attempted to trace the copyright holders of all material reproduced in this publication and apologize to copyright holders if permission to publish in this form has not been obtained. If any copyright material has not been acknowledged please write and let us know so we may rectify in any future reprint.

Except as permitted under U.S. Copyright Law, no part of this book may be reprinted, reproduced, transmitted, or utilized in any form by any electronic, mechanical, or other means, now known or hereafter invented, including photocopying, microfilming, and recording, or in any information storage or retrieval system, without written permission from the publishers.

For permission to photocopy or use material electronically from this work, please access www.copyright.com (http://www.copyright.com/) or contact the Copyright Clearance Center, Inc. (CCC), 222 Rosewood Drive, Danvers, MA 01923, 978-750-8400. CCC is a not-for-profit organization that provides licenses and registration for a variety of users. For organizations that have been granted a photocopy license by the CCC, a separate system of payment has been arranged.

Trademark Notice: Product or corporate names may be trademarks or registered trademarks, and are used only for identification and explanation without intent to infringe.

Library of Congress Cataloging-in-Publication Data

Names: Allamaraju, Subrahmanyam, author. | Prasada Rao, Ganti, 1942- author.
Title: Identification of continuous-time systems : linear and robust parameter estimation / Allamaraju Subrahmanyam and Ganti Prasada Rao.
Description: First edition. | New York, N.Y. : CRC Press/Taylor & Francis Group, 2020. | Series: Engineering systems and sustainability | Includes bibliographical references and index.
Identifiers: LCCN 2019032828 (print) | LCCN 2019032829 (ebook) | ISBN 9780367371432 (hardback) | ISBN 9780429352850 (ebook)
Subjects: LCSH: Automatic control--Mathematical models. | Linear time invariant systems. | System identification. | Parameter estimation.
Classification: LCC TJ212.2 .A44 2020 (print) | LCC TJ212.2 (ebook) | DDC 629.8/95--dc23
LC record available at https://lccn.loc.gov/2019032828
LC ebook record available at https://lccn.loc.gov/2019032829

Visit the Taylor & Francis Web site at
http://www.taylorandfrancis.com

and the CRC Press Web site at
http://www.crcpress.com

Contents

List of Figures

List of Tables

Preface

About fifty years ago in a survey paper (*Automatica* 1971, 123–162) Åström and Eykhoff remarked that the field of System Identification at that time was like a bag of tricks having no organization. That bag was also full of methods only for discrete time (DT) models. Hardly any attention has been given to continuous-time (CT) model (CTM) identification despite the natural relevance to physical systems and the role of CTMs in our understanding of the control of dynamical systems. When required, CTMs were obtained via DT models in an "indirect approach." In subsequent years, methods in a "direct approach" for CTM identification were reported and found to outperform the indirect methods (Ljung 2003). The primary objective of the direct approaches is to retain the system parameters as in their native CTMs. Direct methods are characterized by devices to avoid numerical computation of derivatives of process signals for use in the parameter estimation algorithms, which are also interpreted as some kind of prefilters to provide smoothed differentiated input-output data for parameter estimation. Thus, the direct methods of identifying CTMs are viewed as methods comprising two stages: the prefiltering primary stage, followed by the secondary stage of parameter estimation. Thus, the derivative measurement problem in CTMs is resolved by diverse methods in the primary stage unified into a linear dynamic operator (LDO) or prefilter stage. In the direct approach, the physically meaningful parameters in the CT domain are retained and estimated. The secondary stage—parameter estimation—is standard with the existing statistical methods. Subsequent efforts to handle "uncertainties" and "deficiencies in the knowledge about systems" resulted in the application of robust control methods. Uncertainties are attributed to "noise and disturbances" and "deficiencies" to unmodeled dynamics and handled by $\mathcal{H}_\infty$ methods. Thus, in identifying CT systems, the issues on the side of the signals were handled through the LDO or prefilters to preserve the system parameters in rational transfer function models in the CT domain. On the side of the models, methods consider models that are linear in unknown parameters in order to render the parameter estimation linear.

Several papers surveying direct approaches to CTM identification have appeared over the last four decades—the most recent by Rao and Unbehauen (2006). Among the books that are available on direct approaches, some are devoted to orthogonal function techniques (Rao 1983; Jiang and Schaufelberger 1992; Datta and Mohan 1994). In their book Unbehauen and Rao (1987) presented for the first time LDO techniques of CT identification in a unified way together with the so-called adaptive model methods. Apart from these, some books and special journal issues presenting collections of diverse works have also appeared.

This book is devoted to the identification of CTMs that are linear in their parameters, starting with Markov parameter (MP) and time-moment models. Other linear-in-parameter models—transfer functions represented as Poisson, Lauerre, and Kautz series—are also considered. To enhance the secondary stage of CT system parameter estimation, linear and robust estimation aspects are detailed.

The term "robust" is used in various contexts to denote various "robust" attributes in systems and control. In this book, this word is used in two contexts: when the "estimation (algorithm) is robust," and when "the estimate is robust." The former is about "rejection" (by the algorithm) of errors, while the latter is about "validity" (of the estimate) for a given or predicted bounded set of realizations of errors. Here, "errors" are the effects of the "unmeasurable" stochastic/deterministic disturbances not influenced by the system excitation) and the "unmodeled" (due to model simplification) disturbances on the system output.

Among the various possible parameterizations of models for parametric identification, a class of parameterizations that are linear leads to linear estimation algorithms (e.g., simple least squares estimation) that are robust to zero-mean disturbances. Here the estimates will converge asymptotically to the same set of values as will be obtained in the disturbance-free case, for all possible realizations of zero-mean disturbances. This book attempts to exploit this well-known fact for continuous system identification. Linear-in-parameter models with MPs and time moments (TMs) are the subjects of a study in this book in Chapters 2 and 3.

To ensure validity of the estimate, what is required is a "set" estimate rather than a "point" estimate. Here, given a bounded set of realizations of errors, the estimate is characterized by a bounded set, and the occurrence of any particular value is equally probable among such a set. Set estimation is thus a deterministic counterpart of variance estimation. "Set-membership identification" (SMI), a well-developed branch of system identification, deals with this problem. This problem has been of considerable interest in the past due to its relevance to robust control, where, for robust controller design, such parameter (or equivalently, transfer function) sets are assumed to be available *a priori*. This problem is more involved when the available priors on errors are not in a form suitable for processing by SMI techniques. A number of approaches have been applied toward solving this problem. This activity is appropriately called "error quantification," and it is taken up for investigation in Chapters 4 and 5 of this book. Such investigations are of importance by themselves toward assessing the quality of the estimated model, with a measure of the size of the estimated set indicating the quality.

The following is the outline and organization of this book.

Chapter 1 provides a brief overview of the field of CT system identification and sets the stage for the presentation of estimation of linear-in-parameter models of CT systems.

Chapter 2 presents an MPs approach to CT system identification. Reference to CT MPs is rare in the field of CTM identification, whereas reference to CT MPs is frequently made in the related field of model order reduction. This scarcity is justifiable in view of the natural but difficult-to-estimate form in which MPs are defined for CT systems [see Eq. (1.14)]. Dhawan et al. (1991) made the first attempt in this direction and pointed out some of the implications of using MPs for CT single-input single-output (SISO) system identification, along with some interesting results. Motivated by these results, in this chapter, the problem of irreducible CTM identification via MP estimation for multi-input multi-output (MIMO) systems is considered.

Here, the basic definition of MPs is generalized to make the model parameterization more flexible. However, due to truncation of the infinite length MP sequence (MPS), an undermodeling error is inherent. For the case of finite dimensional systems, a simple iterative but time recursive algorithm employing the pole-placement technique called the "bias reparameterization algorithm" is proposed to remove this undermodeling error in order to finally give a rational transfer function matrix estimate. This algorithm is then analyzed for its convergence. Comparisons with some of the existing methods are made via numerical examples to establish the merits of the proposed class of models.

Chapter 3 considers CTMs parameterized with TMs of system impulse response function. These "not-directly realizable" models have never before attracted the attention of system identification researchers. This chapter attempts to bring these models into the mainstream of model parameterizations via basis functions. Most of the contents of this chapter run in parallel to those of Chapter 2. This is because of the structural similarities between MP models and TM models, although these two have altogether different origins. Given that the basis for TM models is in terms of differentiators, Poisson filtering is introduced to obtain a new realizable basis that is near-orthogonal. A variant of the bias reparameterization algorithm is also suggested here to tackle the inherent undermodeling error.

The two sets of basis functions presented in this and the previous chapters are also compared for their performance, and some guidelines are given for approximating complex systems.

Chapter 4 presents a methodology to robustify the celebrated recursive least squares (RLS) algorithm for parameter estimation of linear regression models, via an $\mathcal{H}_\infty$-norm bounded recursive least squares (NBLS) algorithm. This algorithm is a special case of the $\mathcal{H}_\infty$ filter algorithm (Nagpal and Khargonekar 1991; Shaked and Theodor 1992). This algorithm guarantees estimates with the smallest possible estimation error energy, over all possible modeling errors of fixed energy. Governed by a robust criterion function, the proposed recursive algorithm makes cautious updates. Various similarities with the ordinary RLS algorithm are pointed out. Sufficient conditions for the convergence of the proposed estimator are given, and it is shown that, in the asymptotic case, the estimates converge to the RLS solution. Given prior bounds on modeling errors, formulae are derived to compute ellipsoidal parameter error bounds, thus providing for deterministic robust estimation. Such formulae are useful during the exercise of error quantification.

An important consequence of robustification is an active and accelerated estimation. This feature is useful when the available measurement records are short.

Chapter 5 discusses error quantification—a scientific inference problem in which prior assumptions on modeling errors play a very crucial role. It is argued in this chapter that such prior assumptions should be noninformative so as to let the measurement data speak of the system under identification experiment. A critical survey of the existing literature on this problem reveals that, to quantify error, most of the techniques make use of informative priors, which are difficult to realize in practice. The aim of this chapter is to show that even in the absence of informative priors it is possible to quantify error, and a simple off-line error quantification scheme is suggested for this purpose.

Chapter 6 reviews the work in the preceding chapters and provides some directions for further research.

It is hoped that this modest volume will be a useful addition to the literature on identifying CT systems.

Allamaraju Subrahmanyam

Ganti Prasada Rao

Acknowledgments

Much of the work covered in this book was carried out in the excellent academic atmosphere of the Department of Electrical Engineering, Indian Institute of Technology (IIT), Kharagpur, India, and partly at the International Foundation for Water Science and Technology (IFFWSAT), Abu Dhabi.

We are grateful to many of our colleagues at IIT Kharagpur. In particular, we thank Professors Dines Chandra Saha, Amit Patra, and Siddhartha Mukhopadhyay, who gave their valuable time for many sessions of interesting and fruitful discussions. We are also grateful to Professors K.B. Datta, S. Sinha, and other faculty members of the Department of Electrical Engineering for their encouragement.

We are extremely grateful to Dr. Darwish M.K. Al Gobaisi, director of the International Centre for Water and Energy Systems (ICWES), erstwhile IFFWSAT, and past director general of the Water and Electricity Department (WED), Government of Abu Dhabi, for the unusual and generous support that helped us to work together in Abu Dhabi for some time away from IIT Kharagpur and enabled us to participate in several International Federation of Automatic Control (IFAC) conferences to share our ideas with colleagues from many parts of the world.

We thank Professor Graham Goodwin, Australian Laureate Professor of Electrical Engineering at the University of Newcastle, Australia, for his valuable comments on the first draft of the manuscript.

Professor H. Unbehauen, Lehrstuhl Für Elektrische Steurung und Regelung, Ruhr Universität Bochum, Germany, who collaborated with our team at IIT Kharagpur for nearly four decades in the field of identifying continuous-time systems, passed away on May 1, 2019. He was also one of the coeditors of this book series. We miss him very much.

We thank our families for their patience and support.

Allamaraju Subrahmanyam

Ganti Prasada Rao

Authors

Allamaraju Subrahmanyam (aka **Subbu Allamaraju**) received the BTech degree in Electrical and Electronics Engineering from the Jawaharlal Nehru Technological University College of Engineering, Kakinada, India, in 1990. He received the MTech degree in Control Systems Engineering in 1992 and the PhD degree in 1995, both from the Indian Institute of Technology (IIT), Kharagpur, India.

Subbu is currently a vice president of technology at the Expedia Group, based in Seattle, Washington, USA, where he leads several technology initiatives. His current areas of work include cloud computing, distributed systems, and automation of large-scale systems. He is a frequent speaker at various technology conferences on these topics. He also volunteers his time as an expert-in-residence at RippleWorks, providing guidance for nonprofits and social ventures.

Prior to his current role, Subbu was a distinguished engineer at eBay Inc. for five years, where he helped build eBay's private cloud platform across multiple data centers. He also served in various software engineering roles in other technology companies like BEA Systems (now part of Oracle Corporation), Yahoo Inc., Wipro Infotech, and Computervision (now part of Parametric Technology Corporation).

Subbu has published four books over the many years of his work in the United States:

- *RESTful Web Services Cookbook*, O'Reilly Media (March 2010)
- *Professional Java Servlets 2.3* (January 2002)
- *Professional Java Server Programming J2EE*, 1.3 Edition, Wrox (August 2001)
- *Professional Java E-Commerce*, Wrox (January 2001).

Earlier, he published several research papers on the identification of continuous-time systems using Markov parameters and time moments.

Ganti Prasada Rao received the BE degree in Electrical Engineering from Andhra University, India, in 1963 and the MTech (Control Systems Engineering) and PhD degrees in Electrical Engineering in 1965 and 1969, respectively, both from IIT Kharagpur, India. From July 1969 to October 1971, he was with the Department of Electrical Engineering, PSG College of Technology, Coimbatore, India, as an assistant professor. In October 1971 he joined the Department of Electrical Engineering, IIT Kharagpur, as an assistant professor and was a professor there from May 1978 to June 1997. From May 1978 to August 1980, he was the chairman of the Curriculum Development

Cell (Electrical Engineering) established by the Government of India at IIT Kharagpur. From October 1975 to July 1976, he was with the Control Systems Centre, University of Manchester Institute of Science and Technology (UMIST), Manchester, UK, as a Commonwealth Postdoctoral Research Fellow. During 1982–1983, 1985, 1991, 2003, 2004, 2007, and 2009, he visited the Lehrstuhl für Elektrische Steuerung und Regelung, Ruhr-Universität Bochum, Germany, as a research fellow of the Alexander von Humboldt Foundation. He visited the Fraunhofer Institut für Rechnerarchitectkur und Softwaretchnik (FIRST) Berlin, in 2003, 2004, 2007, 2009, and 2011. He was a visiting professor in 2003 at University Henri Poincare, Nancy, France, and a Royal Society-sponsored visiting professor at Brunel University, UK, in 2007. During 1992–1996, he was a scientific advisor to the Directorate of Power and Desalination Plants, Water and Electricity Department, Government of Abu Dhabi and the International Foundation for Water Science and Technology, where he worked in the field of desalination plant control. At present, he is a member of the United Nations Educational, Scientific and Cultural Organization Encyclopedia of Life Support Systems (UNESCO-EOLSS) Joint Committee.

He has authored/coauthored four books: *Piecewise Constant Orthogonal Functions and Their Applications to Systems and Control*; *Identification of Continuous Dynamical Systems—The Poisson Moment Functional (PMF) Approach* (with D.C. Saha); *General Hybrid Orthogonal Functions and Their Applications in Systems and Control* (with A. Patra), all three published by Springer in 1983, 1983, and 1996, respectively; and *Identification of Continuous Systems* (with H. Unbehauen), published by North Holland in 1987. He has also coedited (with N.K. Sinha) *Identification of Continuous Systems—Methodology and Computer Implementation*, published by Kluwer in 1991. He has authored/coauthored over 150 research papers.

He is/was on the editorial boards of the *International Journal of Modeling and Simulation, Control Theory and Advanced Technology* (C-TAT), *Systems Science* (Poland), *Systems Analysis Modeling and Simulation (SAMS)*, the *International Journal of Advances in Systems Science and Applications* (*IJASSA*), and the *Students' Journal of IETE* (India). He was a guest editor of three special issues: one of *C-TAT on Identification and Adaptive Control—Continuous Time Approaches*, Vol. 9, No. 1, March 1993, and the *Students' Journal of IETE* on Control, Vols. I and II, 1992–1993, *System Science, Special Issue dedicated to Z. Bubnicki*, Vol. 33, No. 3, 2007. He organized several invited sessions in International Federation of Automatic Control (IFAC) symposia on identification and system parameter estimation in 1988, 1991, 1994 and in World Congress in 1993. He was a member of the international program committees of these symposia during 1988–1997. He was a member of the IFAC Technical Committee on Modelling, Identification and Signal Processing in 1996. He was chairman of the Technical Committee of the 1989 National Systems Conference in India. He is coeditor (with A. Sydow) of the book series "Numerical Insights Series" published by Taylor & Francis Group. He is a member of the international advisory boards of the *International Institute of General Systems Science* (IGSS), the *Systems Science* Journal (Poland), and the International Congresses of *World Organisation of Systems and Cybernetics* (*WOSC*).

Since 1996, Prof. Rao has been closely associated with the development of the Encyclopedia of Desalination and Water Resources (DESWARE) (www.desware.net) and the EOLSS, developed under the auspices of UNESCO (www.eolss.net).

He has received several academic awards, including the IIT Kharagpur Silver Jubilee Research Award 1985, the Systems Society of India Award 1989, the International Desalination Association Best Paper Award 1995, and an honorary professorship of the East China University of Science and Technology, Shanghai. He was elected to the Fellowship of the Institute of Electrical and Electronics Engineers (IEEE) with the citation "for development of continuous time identification techniques." The International Foundation for Water Science and Technology established the "Systems and Information Laboratory" in the Electrical Engineering Department at IIT Kharagpur in his honor. He was honored by his alma mater—Indian Institute of Technology, Kharagpur, with its Distinguished Alumnus Award in August 2019.

He is listed in several biographical volumes and is a Life Fellow of the Institution of Engineers (India), a Life Member of the Systems Society of India, a Member of the Indian Society for Technical Education, a Fellow of the Institution of Electronics and Telecommunication Engineers (India), a Life Fellow of the IEEE (USA), and a Fellow of the Indian National Academy of Engineering.

List of Abbreviations

BCLS	bias compensating least squares
BPF	block-pulse functions
CD	common denominator
CONTSID	continuous-time system identification
CRAN	Le Centre de Recherche en Automatique de Nancy
CT	continuous time
CTAT	control theory and advanced technology
CTM	continuous-time model
DT	discrete time
DU	Diekmann-Unbehauen
EE	equation error
EOB	ellipsoidal outer bounding
FIR	finite impulse response
GHOF	general hybrid orthogonal functions
GLS	generalized least squares
HF	Haar functions
IET CTA	Institution of Engineering and Technology [formerly, the Institution of Electrical Engineers (IEE)] Control Theory and Applications
IJC	International Journal of Control
IV	instrumental variables
LDO	linear dynamic operation
LS	least squares
LTI	linear time-invariant
MIMO	multi-input multi-output
MISO	multi-input single-output
MP	Markov parameter
MPPS	Markov-Poisson parameter sequence
MPS	Markov parameter sequence
NBLS	norm bounded least squares
NLMS	normalized least mean squares
OE	output error
PCBF	piecewise constant basis functions
PF	Poisson filter
PFC	Poisson filter chain
PMF	Poisson moment functional
PRBS	pseudo random binary sequence
RDU	Rao-Diekmann-Unbehauen
REN	relative error norm

RLS	recursive least squares
SISO	single-input single-output
SMI	set membership identification
SVF	state-variable filters
TFM	transfer function matrix
TM	time moment
TMS	time-moment sequence
WF	Walsh functions
ZOC	zone of convergence

1

Introduction and Overview

1.1 Background

A system is a unified collection of entities or objects called subsystems. A system may occur naturally or be human engineered for a purpose. The objects are naturally interrelated or artificially made to interact. The collection functions together as a whole within certain spatial and temporal boundaries. A system is characterized by its structure, function, and environment. A dynamical system is a system whose state can change in time and space either spontaneously or by the action of external influence.

Dynamical systems are represented by their mathematical models—difference equations or differential equations. Differential equation models arise mostly from physical systems because the governing laws are basically formulated in continuous time (CT). Examples include Newton's laws of mechanics and Faraday's laws of electromagnetism. When a dynamic system is studied intermittently—in discrete time (DT)—the mathematical models are in the form of difference equations. CT and DT descriptions can be transformed into other domains for operational or computational convenience.

The behavior of a system is exhibited in its response to external inputs and is studied by solving the system model differential/difference equations with forcing functions. The reverse of such study is System Identification whose objective is to determine the dynamical system model that best fits the behavior under appropriate observations. System identification involves consideration of a set of (measurable) signals, a set of models, and a criterion (Zadeh 1962). System identification is then concerned with the determination of an appropriate (based on a criterion) mathematical model to describe the input–output behavior of the system under test on the basis of a given set of measured input and output signals and prior information.

For parametric identification of CT systems, an appropriate parameterization of the chosen model structure in a realistic time domain is crucial. The choice of model structure is governed—first by the nature of the system under test, and then by the intended application of the model.

The field of system identification has been surveyed in the past in general, both in CT and DT (Aström and Eykhoff 1971; Niederlinski and Hajdasinski 1979; Billings 1980; Young 1981; Unbehauen and Rao 1990, 1998; Pintelon et al. 1994). These include surveys of linear time-invariant (LTI) lumped dynamical stable CT systems (Young 1981; Unbehauen and Rao 1990, 1998; Pintelon et al. 1994). There are books dealing with CT identification of systems (Rao 1983; Saha and Rao 1983; Unbehauen and Rao 1987) and edited volumes of collected works of different authors (Sinha and Rao 1991; Garnier et al. 2001, 2008; Garnier and Young 2014) on the subject. Special issues of journals were devoted to CT approaches to identification and control (Rao and Unbehauen 1993; Young and Garnier 2014).

This book focuses on the identification of LTI lumped dynamical stable systems with CT processing of input–output signals; and hence, in the sequel we will consider LTI parametric transfer functions for modeling such systems.

CT systems have traditionally been modeled by linear differential equations in the time domain or by rational transfer functions in the complex variable "s" in the frequency domain. However, such models cannot be processed on digital computers without discretizing CT operators and process signals. Consequently, CT systems began to be modeled by difference equations in the time domain or by rational transfer functions in the complex variable "z" in the frequency domain. These models remained popular for their simplicity and the availability of analytical tools until several researchers pointed out the associated drawbacks in the context of identification and control (Gawthrop 1987; Middleton and Goodwin 1990; Mukhopadhyay 1990; Sinha and Rao 1991) such as

- High coefficient sensitivity of these models at fast sampling, leading to problems in parameter estimation and control
- Clustering of poles near (1,0) point in the z-plane at fast sampling leading to undesirable numerical ill-conditioning
- Appearance of non-minimum phase zeros due to sampling
- Lack of uniqueness of models due to their dependence on sampling rate
- Loss of physical significance of parameters due to discretization even for partially known CT systems
- Loss of prior knowledge available in the CT model.

See Mukhopadhyay et al. (1992) and the references therein for details. Further research to avoid such drawbacks has led to the currently popular approximate CT modeling in which the CT operators and signals are DT approximated while retaining the CT model parameters. This approach to CT model (CTM) identification has gained a lot of attention in recent years (Unbehauen and Rao 1987; Sinha and Rao 1991). In this approach, the DT (sampled or averaged) measures of the process signals are realized using sampled or averaged measurements, and the filtered versions of signals that arise in CT modeling are computed via DT approximations of the CT differentiation operator $p\left(=\frac{d}{dt}\right)$. An important feature of this modeling is that the resulting discrete domain is in harmony with the CT s-domain, approaching it with rapid sampling.

1.2 Introduction

Let $\mathcal{G}$ be an LTI-stable dynamical system with an unknown transfer function $G^0(p)$ defined as

$$y(t) = G^0(p)u(t) + v(t), \tag{1.1}$$

where $v(t)$ is an unknown disturbance term. Define the averaged/sampled (at averaging/sampling interval T_s) measurement data set

$$\mathbf{Y}^N = \{u(k),\ y(k),\ k = 1,\ \ldots, N\}. \tag{1.2}$$

The identification problem considered here is that, given $\mathbf{Y}^N$, along with some prior knowledge concerning the dynamics of $\mathcal{G}$ and the input–output signals, to compute a nominal transfer function estimate $G(\overline{p},\theta)$ to describe the dynamics of $\mathcal{G}$ as closely as possible by minimizing some suitable norm of modeling errors, and to make a quantitative statement assessing the quality of the estimate. Here $\overline{p}$ is a DT approximation (such as obtained by forward or backward differencing, or trapezoidal rule) of the CT differentiation operator p, and θ is the unknown parameter vector.

In terms of the parameterized model structure $G(\overline{p},\theta)$,

$$y(k) = G(\overline{p},\theta)u(k) + e(k), \tag{1.3}$$

where $e(k)$ is the total modeling error that includes effects of unmodeled dynamics due to restricted complexity modeling, unmeasurable disturbances $v(k)$ and possibly effects of unknown initial conditions.

Remark 1.1 Assumptions on $v(k)$

In conventional system identification, disturbances and errors are treated to be realizations of stochastic processes, and $v(k)$ is generally assumed to be a realization of a zero-mean stationary stochastic process independent of $u(k)$.

On the other hand, in the case of bounded-error/worst-case identification (for an overview, see Norton 1987a; Deller 1989; Arruda and Favier 1991; Milanese and

Vicino 1991; Wahlberg and Ljung 1992) where disturbances and errors are treated in a deterministic framework, an assumption on $v(k)$ in terms a bound C_v, is

$$v(k)^2 \leq C_v < \infty. \tag{1.4}$$

In addition, the following assumptions are generally imposed on $v(k)$,

$$\lim_{N\to\infty} \frac{1}{N}\sum_{k=1}^{N} v(k) = 0,$$

$$\lim_{N\to\infty} \frac{1}{N}\sum_{k=1}^{N} v(k)u(k-\kappa) = 0, \ \forall\kappa,$$

and

$$\infty > \lim_{N\to\infty} \frac{1}{N}\sum_{k=1}^{N} [v(k)]^2 = C > 0$$

1.3 Role of Model Parameterizations in System Identification

Although linearity of models is primarily governed by the nature of the system under test and the intended model application, there is another kind of linearity of importance, namely, between the dependent variables and the parameters, that is, linearity of the model parameterization. A parameterization is linear (or in other words, a model structure is linear in parameters) if second- and higher-order partial derivatives of $G(p,\theta)$ with respect to $\boldsymbol{\theta}$ vanish for all $\boldsymbol{\theta}$. Note that linearity of a parameterization is different from the linearity of the model in terms of input–output behavior. Even nonlinear models can be linearly parameterized.

To appreciate the importance of linearly parameterized models in identification, consider, for example, a DT model. Its representation as a rational transfer function leads to nonlinear estimation, whereas a finite impulse response (FIR) model representation leads to simple linear estimation problems. However, in various problems of control, signal processing and identification, use of rational transfer functions or differential/difference equations as models has been very popular and a majority of theoretical contributions to these fields are based on such models.

Let $G(\overline{p},\theta)$ be parameterized as a proper rational transfer function

$$G(\overline{p},\theta) = \frac{B(\overline{p},\theta)}{A(\overline{p},\theta)}, \tag{1.5}$$

where B and A are polynomials in $\bar{p}$. Then the model output error (OE) in sampled form is

$$\epsilon_{\mathrm{OE}}(k) = y(k) - \frac{B(\bar{p},\theta)}{A(\bar{p},\theta)} u(k) \tag{1.6}$$

and a suitable parameter estimation criterion is to minimize

$$J_{\mathrm{OE}} = \sum_{k=1}^{N} \epsilon_{\mathrm{OE}}(k)^2. \tag{1.7}$$

Given that Eq. (1.6) is nonlinear in parameters, minimization of the above criterion function requires nonlinear optimization. To simplify matters, most of the identification approaches consider the equation error (EE)

$$\epsilon_{\mathrm{EE}}(k) = \frac{A(\bar{p},\theta)}{E(\bar{p},\theta)} y(k) - \frac{B(\bar{p},\theta)}{E(\bar{p},\theta)} u(k) \tag{1.8}$$

and a criterion

$$J_{\mathrm{EE}} = \sum_{k=1}^{N} \epsilon_{\mathrm{EE}}(k)^2. \tag{1.9}$$

Here $\frac{1}{E(\bar{p})}$ is a linear-dynamic operator of a suitable order that removes the need for direct differentiation of process data (Unbehauen and Rao 1990). These operators also serve the purpose of prefilters used for removing unimportant frequencies form the process data. The prefilter concept originates from the work of Shinbrot (1957), who used a set of modulating functions on the input–output signals mainly to shift the derivative operations from process signals on to the chosen modulating functions whose derivatives need not be obtained by undesirable numerical computations. The preservation of the parameters in their CT form is desirable for many reasons, although some suggest an approach to identify CT systems by an indirect method in which a DT model is obtained and then transformed into CT form. This is not without difficulties given that a DT model obtained by discretizing a CT model is unique and going back to its native CT version is problematic. Therefore, direct methods of identifying CT models are preferred. In the direct approach to CT identification, the input and output signals are prefiltered, as mentioned before, in several ways that are connected to the modulating function method. Unbehauen and Rao (1990) in their surveys call this prefiltering a linear dynamic operation (LDO), which views a modified version of the modulating function method as one using convolution involving a "modulating function" arising out of the impulse response function of a filter.

1.3.1 Poisson Moment Functional Approach

The need to compute the integrals of the products of the signals with modulating functions was later eliminated by choosing exponential impulse response function of a first-order system characterized by the transfer function $1/(s+\lambda)$, $\lambda \geq 0$. This is known as a Poisson filter. Because the pole is far into the left half plane, its transients decay fast and, therefore, the initial conditions are not considered in the convolution integral that gives the output of the filter element. The input and output signals are passed through chains of such elements called Poisson filter chains (PFCs).

Given that $\lambda \geq 0$, the bandwidth of the Poisson filter is large and much of the noise passes into the measured outputs of the filter chains. With $\lambda = 0$, the filter element becomes an integrator that can filter high-frequency components of the noise spectrum. Parameter estimation can also be done in the presence of unknown initial conditions. The output signals of the various stages of the Poisson filter chain are known as Poisson moment functionals (PMFs). Much work is reported on the use of PMFs in identifying CT systems (Fairman and Shen 1970; Fairman 1971; Saha and Rao 1980a,b, 1981, 1982; Saha et al. 1982). In the book by Saha and Rao (1983), the authors comprehensively cover the use of PMFs in CT system identification. Rao and Sivakumar (1976) applied the PMF method also to determine an unknown time delay in a system. In 1982, Rao and Sivakumar processed the PMF through a Kalman filter and obtained good results on a simulated system. Bapat (1993) made some extensions of the PMF method. Garnier and colleagues reported several results during 1994–1995 on the application of PMF method in system identification.

With $\lambda = 0$, the Poisson filter element becomes a time integrator. Methods related to this case of PMF were first reported by Diamessis (1965a, 1965b, 1965c), who applied the repeated integration method to LTI, time-varying, and nonlinear systems. Sagara and Zhao (1988, 1989, 1991) used integrating filters for online identification of CT systems.

1.3.2 Integral Equation Approach

With repeated integration, the model differential equation becomes an integral equation in which the integral operations on process input and output signals can be safely performed, resulting in filtering noise. The integral operation in the integral equations is realized as an algebraic operation through an operational matrix for integration if the signals are characterized by orthogonal basis functions in the time domain. Several classes of orthogonal functions such as continuous basis functions (Sinusoidal, Jacobi, Chebychev, Laguerre functions) and piecewise constant basis functions (PCBFs) [Walsh functions (WFs), Haar functions (HFs), and block-pulse functions (BPFs), collectively called PCBFs] or general hybrid orthogonal functions (GHOFs) (Patra and Rao 1996) can be used.

BPFs, being disjoint and simple in their form, render the algorithms free from matrix operations and give rise to recursive solutions. The "spectral components" are used as filtered data in place of sampled data. Rao (1983) presented a comprehensive account of PCBFs, and Jiang and Schufelberger (1992) did the same with BPFs. Datta and Mohan (1994), on the other hand, focused on continuous basis functions. The signals so obtained by a prefiltering operation—which is termed the "primary stage" by Unbehauen and Rao in their surveys—can be sampled for use in standard estimation algorithms that are essentially the same as in DT model estimation using samples of raw unfiltered input–output signals. Rao and Unbehauen (2006) presented a comprehensive survey of CT identification. Figures 1.1 and 1.2 show the indirect and direct methods of identification for CT systems. Figure 1.3 shows the Poisson Filter Chain, which becomes a chain of integrators when $\lambda = 0$.

Padilla et al. (2015) presented the CONTSID Toolbox compatible with MATLAB, which has been developed at CRAN over the many years since the 1990s.

Because Eq. (1.8) is linear in parameters, the parameter estimation is simplified to the linear recursive least squares (RLS) estimation. However, EE minimization has its own disadvantages.

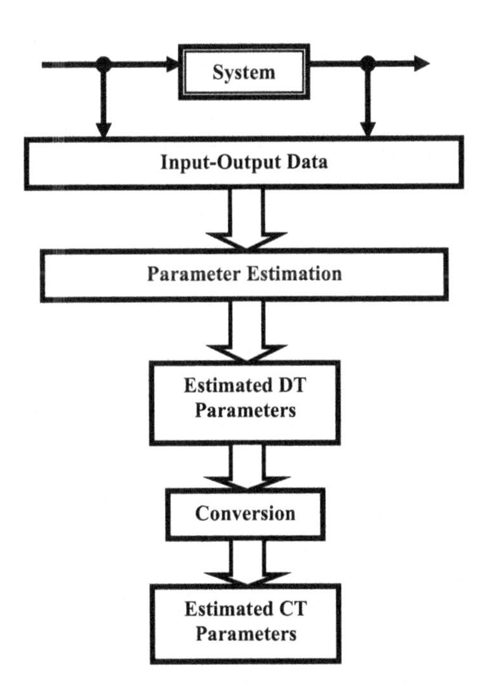

FIGURE 1.1
Indirect identification of CT parameters via DT parameters.

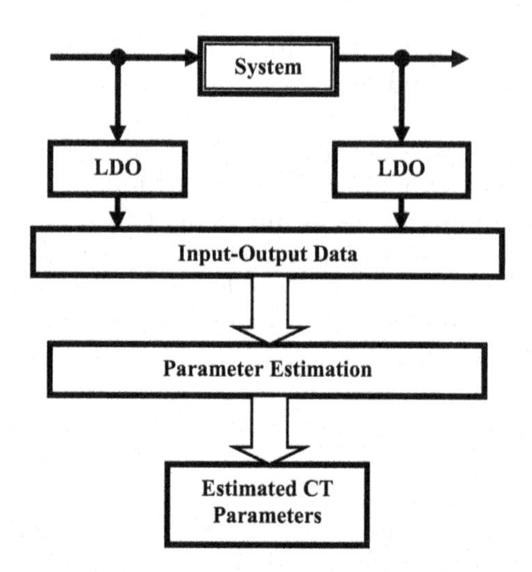

FIGURE 1.2
Direct identification of CT parameters via DT parameters.

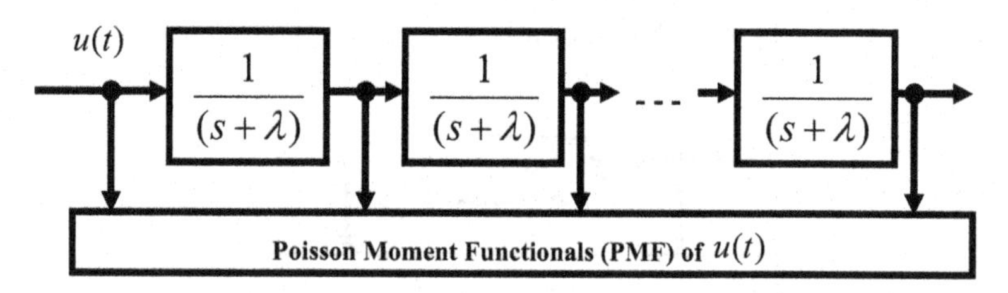

FIGURE 1.3
Poisson filter chain.

1.3.3 Biased Estimation

The parameter estimates will be biased when the EE is not white (see also Remarks 1.2 and 1.3). To tackle this, variants of the ordinary least squares (LS) algorithm like generalized least squares (GLS) and instrumental variables (IVs) have been proposed (Söderström and Stoica 1989). Application of these for CT system identification has been observed to be computationally demanding (Mukhopadhyay et al. 1991). Bias compensating least squares (BCLS) methods have also been proposed (Zhao et al. 1991a, 1991b, 1992; Yang et al. 1993) for this purpose. However, performance of some of these may not be satisfactory when there is significant modeling error.

1.3.4 Reducible Models (for Multivariable Systems)

Another disadvantage is that EE formulation with multivariable transfer function matrix (TFM) model necessitates a canonical form having a least common denominator (CD) of all the elements of the TFM. Due to the CD structure, the size of the unknown parameter vector is inflated and the question of irreducible TFM model identification has been an unsolved problem. To solve this, although partially, the multivariable or the multi-input multi-output (MIMO) TFM model is decomposed into multi-input single-output (MISO) submodels requiring CD only over the corresponding row of the TFM. Based on this, for DT model identification, a two-stage algorithm (called a DU algorithm) was proposed by Diekmann and Unbehauen (1979), whose CT version was later explored by Mukhopadhyay et al. (1991). A Gauss-Seidel–type iterative algorithm, also called a Rao-Diekmann-Unbehauen (RDU) algorithm, was suggested by Rao et al. (1984). Although the RDU algorithm estimates irreducible models, its CT version is computationally demanding due to a large number of simulations of the estimated models to be carried out in the actual implementation. For a brief survey, see Mukhopadhyay and Rao (1991) and Mukhopadhyay et al. (1991).

1.3.5 Distribution of Estimation Errors

Yet another difficulty arises in the case of experiment design for a prescribed distribution of errors over frequencies in restricted complexity model estimation (Wahlberg and Ljung 1986). Using Parseval's theorem, the frequency domain counterpart of the EE criterion [Eq. (1.9)] in the limit as $T_s \to 0$ is

$$J_{EE} = \int_0^\infty \left| \frac{A(j\omega,\theta)}{E(j\omega)} U(j\omega) \right|^2 \left| G^0(j\omega) - \frac{B(j\omega,\theta)}{E(j\omega)} \right|^2 d\omega, \tag{1.10}$$

where $U(j\omega)$ is the Fourier transform of the input. In the above, the first term on the right-hand side is a weighting function that manipulates the second term (error) over frequencies. With the chosen rational transfer function model structure, it is now clear that this weighting function is a function of the yet-unknown $A(\overline{p},\theta)$ thus making on-line experiment design impossible. Off-line design, however, is shown to be possible (Bapat 1993).

Despite the above, nonlinear parameterizations of $G(\overline{p},\theta)$ as in Eq. (1.5) are very popular and widely used for their ability to adequately approximate most of the real-world dynamical processes.

Remark 1.2

In the case of bounded-error/worst-case identification, Fogel and Huang (1982) show that $\epsilon(k)$ should satisfy

$$\lim_{N\to\infty} \frac{1}{N} \sum_{k=1}^{N} \epsilon(k) = 0,$$

$$\infty > \lim_{N\to\infty} \frac{1}{N} \sum_{k=1}^{N} [\epsilon(k)]^2 = C > 0,$$

and

$$\lim_{N\to\infty} \frac{1}{N} \sum_{k=\kappa}^{N} \epsilon(k)\epsilon(k-\kappa) = 0,$$

for consistent estimation. A stochastic interpretation of these deterministic conditions is that $\epsilon(k)$ should be a realization of a stationary, uniformly distributed white noise process.

Remark 1.3

In the case of CTM identification, whiteness of EE is not sufficient for consistency. An additional requirement is that $\bar{p}^{-1}$ (written as a function of the forward shift operator q) should be strictly proper. An example is the δ-operator approximation, for which

$$\bar{p}^{-1} = \frac{T_s}{q-1}.$$

On the other hand, approximations based on backward differences or trapezoidal rule, for which

$$\bar{p}^{-1} = \frac{T_s q}{q-1}$$

and

$$\bar{p}^{-1} = \frac{T_s}{2}\left(\frac{q+1}{q-1}\right),$$

respectively, lead to inconsistent estimation even if the EE happens to be white.

An alternative class of models under a general basis function framework was explored in various problems of identification and control. This is a special class of models in which the model output is formed as a linear combination of certain functions (known as basis functions; see, e.g., Goodwin et al. 1991a) of the input. These models have the following structure:

$$G(\delta,\theta) = \theta^{\mathrm{T}} \mathcal{B}(\delta), \tag{1.11}$$

where $\theta = [\theta_1, \theta_2, \ldots, \theta_d]^{\mathrm{T}}$; $\mathcal{B}(\delta) = [\mathcal{F}_1(\delta), \mathcal{F}_2(\delta), \ldots, \mathcal{F}_d(\delta)]^{\mathrm{T}}$ is the basis; and $\delta = \bar{p}, q^{-1}, s$ or s^{-1}, as the case may be. The performance of these models depends on the shape and span of the chosen basis for a given model order d. For these models, EE and OE are the same.

Models of the above structure evolve very naturally from truncated series expansions of the transfer function. For example, in the DT case, the system transfer function may be written in terms of its impulse response function as

$$G^0(q) = \sum_{i=1}^{\infty} h_i q^{-i},$$

where q^{-1} is the backward shift operator and $\{h_i\}$ is the impulse response sequence. This motivates the following parameterization:

$$G(q, \theta) = \theta^{\mathrm{T}} \mathcal{B}(q^{-1}), \tag{1.12}$$

where $\theta = [h_1, h_2, \ldots, h_d]^{\mathrm{T}}$ and $\mathcal{B}(q^{-1}) = [q^{-1}, q^{-2}, \ldots, q^{-d}]^{\mathrm{T}}$. The quality of this approximation depends on the rate of convergence of the impulse response sequence. Poles of $G^0(z)$ close to the unit circle result in a slow rate of convergence and, consequently, a high model order for a given tolerance. Hence, for rapidly sampled CT systems the rate of convergence of the approximation will be very slow, and in the limit as $T_s \to 0$, the DT poles converge to unity and consequently the approximation fails to converge. Moreover, even for convergent approximations, a high model order is required as the memory of the basis (shift operator) is very short (unity). This motivates alternative model representations in the CT domain that have better convergence properties and are less sensitive to the sampling rate.

In the CT case, the transfer function $G^0(s)$ may be expanded about $s = \infty$ as a complex power series in s^{-1} as

$$G^0(s) = \sum_{i=1}^{\infty} h_i s^{-i},$$

leading to a parameterization

$$G(s, \theta) = \theta^{\mathrm{T}} \mathcal{B}(s^{-1}), \tag{1.13}$$

where $\theta = [h_1, h_2, \ldots, h_d]^{\mathrm{T}}$ and $\mathcal{B}(s) = [s^{-1}, s^{-2}, \ldots, s^{-d}]^{\mathrm{T}}$. It is easy to verify that $\{h_i\}$ are the CT Markov parameters of $G^0(s)$ defined as

$$h_i = \left. \frac{d^{i-1}}{dt^{i-1}} g^0(t) \right|_{t=0} \tag{1.14}$$

where $g^0(t)$ is the impulse response of $G^0(s)$.

Considering a similar expansion of $G^0(s)$ about $s = 0$, we get models parameterized in terms of normalized time moments of the impulse response $g^0(t)$ of $G^0(s)$, that is,

$$G(s,\theta) = \theta^T \mathcal{B}(s), \tag{1.15}$$

where

$$\theta = [m_1, m_2, \ldots, m_d]^T,$$

$$\mathcal{B}(s) = \left[s^1, s^2, \ldots, s^d\right]^T,$$

and

$$m_i = \frac{(-1)^i}{i!} \int_0^\infty t^i g^0(t) dt \tag{1.16}$$

are the normalized time moments.

Various other choices of basis functions are also possible. Very popular among these are Laguerre and Kautz filters. Laguerre filters form a basis,

$$\mathcal{B}_{\text{LAG}} = \left[\frac{1}{p+\lambda}, \frac{1}{p+\lambda}\left(\frac{p-\lambda}{p+\lambda}\right), \ldots, \frac{1}{p+\lambda}\left(\frac{p-\lambda}{p+\lambda}\right)^{-1}\right]^T, \tag{1.17}$$

with $\lambda > 0$, and Kautz filters form

$$\mathcal{B}_{\text{KAUTZ}} = [\psi_1(s), \psi_2(s), \ldots, \psi_d(s)]^T, \tag{1.18}$$

where

$$\psi_{2k-1}(s,b,c) = \frac{s}{s^2+bs+c}\left[\frac{s^2-bs+c}{s^2+bs+c}\right]^{k-1},$$

$$\psi_{2k}(s,b,c) = \frac{1}{s^2+bs+c}\left[\frac{s^2-bs+c}{s^2+bs+c}\right]^{k-1},$$

with $b > 0$, $c > 0$, and $k = 1, 2, \ldots$. For details see Wahlberg (1994) and the references therein.

The roles played by basis functions in continuous and discrete system modeling are discussed by Goodwin et al. (1991a, 1991b). The reason such models are being investigated is that these are linear-in-parameter models with which OE minimization becomes linear. Consequently, the following features arise, some of which are exploited in this book.

- The span of the basis precisely controls the performance of these models. Hence, by an intelligent choice of the basis, even complex transfer functions can be modeled significantly accurately with a small number of parameters. Here prior knowledge of system dynamics may be effectively used in choosing the right basis.
- In the limit as $N \to \infty$, the LS estimate $\hat{\theta}$ in the presence of zero-mean $v(k)$ tends to $\hat{\theta}^*$, where $\hat{\theta}^*$ is the limiting estimate in the absence of $v(k)$. This holds good even for colored $v(k)$. The estimation is thus robust in the sense that the effects of disturbances are rejected as $N \to \infty$.
- With MIMO models, there is no inflation of the unknown parameter vector.
- The weighting function in Eq. (1.10) is a function of the input spectrum and the prefilter (E) only, and thus on-line experiment design for a prescribed error distribution is possible.

Motivated by these, this book, in its first phase, studies CT models parameterized in terms of Markov parameters and time moments for CT MIMO system identification. These models are generalized so as to include prior knowledge of system dynamics and to ensure that even low-order models give an adequate representation of the system under test.

1.4 Error Quantification: An Engineering Necessity

A system identification exercise would be complete when a statement of quality of the estimated model is also added. Error quantification, to assert model quality as probabilistic or deterministic bounds on the total error in the estimated model is an engineering necessity.

In many practical situations where physical phenomena governed by complicated (and sometimes inexplicable) relationships are represented by simplified models, modeling error to some extent is inevitable. The total modeling error in the estimate can be split into errors that are due to these modeling approximations that lead to unmodeled dynamics and those due to noise as well as finite amount of data used in estimation. However, over the more than 40 years of research on system identification, not much attention has been paid to the former. Most of the time, the estimation error has been assumed to be due only to additive stochastic disturbances corrupting the measurements, and asymptotic formulae were provided to compute the variance of the estimates (Ljung 1999; Söderström and Stoica 1989).

There is also a branch of system identification, known as set membership identification (SMI) for deterministic robust estimation, that treats disturbances in a completely deterministic framework assuming that disturbances are unknown but bounded. SMI is also known as "worst-case identification" or "unknown-but-bounded error identification." Surveys on this topic can be found in Milanese and Vicino (1991), Arruda and Favier (1991), and Deller (1989). SMI techniques produce guaranteed error bounds provided the prior information (bounds on disturbances) is correct.

The role of unmodeled dynamics and the importance of quantifying resulting errors in estimation has been recognized for quite some time and some preliminary ideas were suggested (Anderson et al. 1978; Kabaila and Goodwin 1980; Ljung 1987). One of the first results on characterization of estimation errors was due to the work of Wahlberg and Ljung (1986), where formulae that give an idea of the distribution of error in transfer function estimate over frequencies in the case of LS estimation are given. Such formulae are useful in guiding identification experiments for a prescribed distribution of error over frequencies by tuning a set of design variables of the estimator. However, they cannot give any insight into what the bounds on estimation error are. Both the classical stochastic estimation and deterministic robust estimation get complicated in this case because this component of total modeling error does not usually fit the description of disturbances.

On the other hand, advances in robust control design methods in 1980s went independent of system identification theory by assuming that explicit worst-case/deterministic bounds were available on error in transfer function or parameter estimates (or both) (Dorato 1987; Dorato and Yedawalli 1990; Maciejowski 1989), thus making the results that the identification theory had produced incompatible with the tools and assumptions of robust control design. Realization of this incompatibility has spurred active research in the field of system identification leading to the development of a class of control-oriented system identification methods that yield an explicit bound on the resulting identification error (see the special issue of *IEEE Transactions on Automatic Control*, July 1992).

The main focus of research in system identification is now on the estimation of accurate error bounds, whether probabilistic or deterministic (or combined), and novel identification techniques are being produced for the sake of delivering such error bounds. As pointed out by Hjalmarsson (1993), this topic is still in its infancy and there is a certain degree of incoherence between the approaches of different groups of researchers. There is no uniformity either among the various prior assumptions made or the form in which the results are presented. Recent surveys of available approaches can be found in Gevers (1991), Milanese and Vicino (1991), Wahlberg and Ljung

(1992), and Kosut (1993). As will be seen later in this book, most of the available approaches start with information-rich or informative priors that are difficult to realize in practice.

With this background, this book, in its second phase, takes up the problem of deterministic robust estimation for quantifying error in parametric/transfer function estimation as ellipsoidal/elliptical bounds when the available priors are not informative. For this, a robust version of the RLS algorithm for conservative estimation is first developed. Formulae are then derived to compute ellipsoidal outer bounds on estimation error in parameters, given prior bounds in the form of Eq. (1.4) on the total modeling error. However, because such priors are informative and are difficult to obtain in practice, an off-line technique is proposed to actually predict them by modeling.

2

Markov Parameter Models

2.1 Introduction

The use of Markov parameters (MPs) for parameterizing models is not new in the field of system identification. Although not for CTM identification, there have been references to DT MPs for the identification of DT multivariable models. DT models of the form of Eq. (1.12) were considered for identification by Mehra (1971) and Sinha et al. (1978). The problem of direct DT MP estimation based on cross-correlation between the output and a white noise input has also been studied by Sinha et al. (1978). For a survey on related issues, see Niederlinski and Hajdasinski (1979).

However, in the CT situation, the reference to MPs is rare. This is justifiable in view of the natural but difficult-to-compute form in which MPs are defined for CT systems. In this case, MPs are various higher-order derivatives of the impulse response function at time $t = 0$ [Eq. (1.14)].

Dhawan et al. (1991) attempted in this direction and suggested the use of MP models for single-input single-output (SISO) CTM identification. They proposed to realize the MP model (Eq. 1.13) as an integral equation in which the integrals are realized using block-pulse functions (BPFs), thus avoiding the definition of MPs as in Eq. (1.14). However, truncation of the MP model as in Eq. (1.13) often leads to diverging approximations, due to which the estimation may fail to converge. Dhawan et al. (1991) pointed this out and suggested a simple generalization of the original MP model to ensure convergent approximations (see also Küper 1992). Further generalization of the basis leading to flexible and well-behaving approximations was later suggested by Subrahmanyam and Rao (1993). The latter also proposed an iterative algorithm to remove the truncation error from the MP model for the case of finite dimensional systems. These works are limited to the case of SISO systems. The material of the present chapter extends these ideas and is addressed toward estimation of irreducible CT models in their TFM format of MIMO systems via MP estimation.

The organization of this chapter is as follows:

- Section 2.2 begins with the basic definition of MPs of MIMO systems and their realization from state space/TFM descriptions. A generalized parameterization with the so-called Markov-Poisson parameters is described, and some guidelines on the choice of certain structural coefficients of this parameterization are given. The resulting models are then descritized using the BPF approximation.
- Section 2.3 presents an iterative algorithm for the finitization of MPSs. This algorithm, which uses the pole-placement technique, is termed a "bias reparameterization algorithm" for reasons to be discussed later.
- Section 2.4 derives certain conditions on the input signal for the parameters to be identified. Unlike the case of CD MISO models using EE minimization, the suggested linearly parameterized models require more meaningful conditions to be satisfied for identifiability.
- Section 2.5 analyses the proposed bias reparameterization algorithm of Section 2.3 for its convergence.
- Section 2.6 presents some results of numerical experiments while bringing out the features of the current proposals.
- Section 2.7 presents some discussion on the chapter.

2.2 Markov Parameter Models

Consider a linear time-invariant CT MIMO system with ν_i inputs and ν_o outputs,

$$\begin{aligned}\dot{\mathbf{x}}(t) &= \mathbf{A}\mathbf{x}(t) + \mathbf{B}\mathbf{u}(t) \\ \mathbf{y}(t) &= \mathbf{C}\mathbf{x}(t) + \mathbf{v}(t),\end{aligned} \tag{2.1}$$

with the usual notation.

The external frequency domain description of the system is

$$\mathbf{Y}(s) = \mathbf{G}^0(s)\mathbf{U}(s) + \mathbf{V}(s), \tag{2.2}$$

where $\mathbf{Y}(s) = \mathcal{L}\,\mathbf{y}(t)$; $\mathbf{U}(s) = \mathcal{L}\,\mathbf{u}(t)$; $\mathbf{V}(s) = \mathcal{L}\,\mathbf{v}(t)$; $\mathcal{L}$ denotes the Laplace transform; and $\mathbf{G}^0(s)$ is the $\nu_o \times \nu_i$ size TFM

$$\mathbf{G}^0(s) = \{G_{ij}(s);\ i = 1, \ldots, \nu_o;\ j = 1, \ldots, \nu_i\},$$

where each element

$$G_{ij}(s) = \frac{B_{ij}(s)}{A_{ij}(s)} = \frac{b_{1,ij}s^{n_{ij}-1} + \cdots + b_{n_{ij},ij}}{s^{n_{ij}} + a_{1,ij}s^{n_{ij}-1} + a_{1,ij}s^{n_{ij}-1} + \cdots + a_{n_{ij},ij}}$$

describes the relation between the jth input u_j and the ith output y_i.

CT MPs of the above system are the coefficients of the power series

$$\mathbf{G}^0(s) = \sum_{l=1}^{\infty} \mathbf{C}\mathbf{A}^{l-1}\mathbf{B}s^{-l} = \sum_{l=1}^{\infty} \mathbf{H}_l s^{-l}, \tag{2.3}$$

whose ijth element is related to the coefficients of $A_{ij}(s)$ and $B_{ij}(s)$

$$h_{l,ij}(s) = b_{l,ij}(s) - \sum_{r=0}^{l-1} h_{r,ij}(s)a_{l-r,ij}(s), \quad l = 1,2,\cdots n_{ij}, \tag{2.4}$$

$$h_{l+n_{ij},ij}(s) = -\sum_{r=1}^{n_{ij}} n_{ij+l-r,ij}(s)a_{r,ij}(s), \quad l = 1,2,\cdots. \tag{2.5}$$

In terms of $\{\mathbf{H}_l\}$,

$$\mathbf{y}(t) = \sum_{l=1}^{\infty} \mathbf{H}_l \mathbf{u}^l(t) + \mathbf{v}(t), \tag{2.6}$$

where $\mathbf{u}^l(t)$ is the lth integral of $\mathbf{u}(t)$. Assuming absolute convergence of the MPS and thus uniform convergence of partial sums, we have a truncated MP model,

$$\tilde{\mathbf{y}}(t) = \sum_{l=1}^{d} \mathbf{H}_l \mathbf{u}^l(t) = \theta^{\mathrm{T}} \mathcal{B}(p^{-1})\mathbf{u}(t), \tag{2.7}$$

where

$$\theta = \left[\mathbf{H}_1, \mathbf{H}_2, \cdots, \mathbf{H}_d\right]^{\mathrm{T}} \text{ and } \mathcal{B}(p^{-1}) = \left[p^{-1}, p^{-2}, \cdots, p^{-d}\right]^{\mathrm{T}} \text{ is the basis.}$$

2.2.1 Generalizations

The above MP model is valid only when the power series expansion of $\mathbf{G}^0(s)$ is absolutely convergent. Unlike DT MPs, which are nothing but samples of the impulse response function, CT MPs may result in diverging approximations even for stable systems. The sequence $\{\mathbf{H}_l\} = \{\mathbf{C}\mathbf{A}^{l-1}\mathbf{B}\}$ and approximations of Eq. (2.6) are convergent only when $\left|\lambda_i\{\mathbf{A}\}\right| < 1, \forall i$, which is not always guaranteed. In such cases, the radius of convergence of power series of

Eq. (2.3) becomes zero. To ensure absolute convergence, this radius should be infinity. For this, and to increase the rate of convergence of the approximation, Subrahmanyam and Rao (1993) proposed a more general version of MPs—known as Markov-Poisson parameters $\{\bar{\mathbf{H}}_l\}$,

$$\bar{\mathbf{H}}_l = \frac{1}{\beta_c^l}\sum_{i=1}^{l} {}^{l-1}C_{i-1}\lambda^{l-i}\mathbf{H}_i, \quad l = 1,2,\cdots; \quad \beta_c > 1, \tag{2.8}$$

and Eq. (2.3) is rewritten with a new basis $\boldsymbol{\mathcal{B}}_{PF}(s) = \left[\frac{\beta_c}{s+\lambda}, \left(\frac{\beta_c}{s+\lambda}\right)^2, \cdots, \left(\frac{\beta_c}{s+\lambda}\right)^d\right]^T$ as

$$\mathbf{G}^0(s) = \sum_{l=1}^{\infty} \bar{\mathbf{H}}_l \left(\frac{\beta_c}{s+\lambda}\right)^l. \tag{2.9}$$

Note that the elements of $\boldsymbol{\mathcal{B}}_{PF}$ are the well-known Poisson filters (Saha and Rao 1983) of increasing orders.

Implications of the above generalization are as follows:

- Zone of convergence (ZOC): increased to $\left|\lambda_i\{\mathbf{A}\} - \lambda\right| < \beta_c, \forall i$.
- Radius of convergence: infinity.
- Predictive ability: By definition, MP models provide good high frequency predictions. The above generalization improves the low frequency performance as well.

2.2.2 Choice of λ

1. $\lambda > 0$ (leftward shift of the imaginary axis): This choice, originally suggested by Subrahmanyam and Rao (1993), is well suited for overdamped systems with poles that are not very close to the imaginary axis (Figure 2.1a).
2. $\lambda < 0$ (rightward shift of the imaginary axis): Choice 1 cannot work on systems whose poles (real/complex) are arbitrarily close to the imaginary axis. This limitation is avoidable by a rightward shift of the imaginary axis followed by a compression of the s-plane by a large β_c. To illustrate, consider the case of an under-damped second-order system

$$G(s) = \frac{\omega_n^2}{s^2 + 2\zeta\omega_n s + \omega_n^2}$$

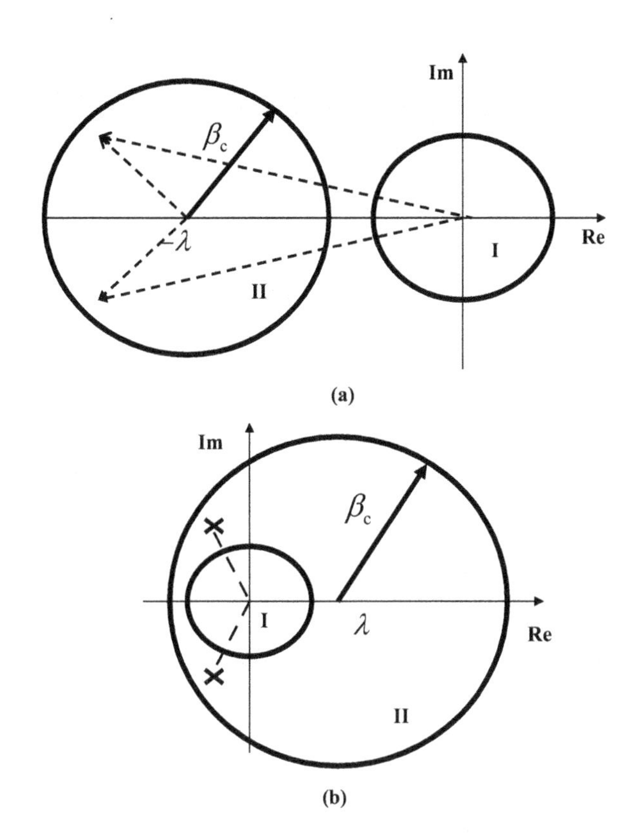

FIGURE 2.1
Expansion of the complex plane: (I) Actual ZOC, (II) Expanded ZOC. (a) $\lambda < 0$ for real and heavily damped poles; and (b) $\lambda > 0$ for lightly damped and imaginary pole.

whose MPS is

$$h_l = \frac{(-\omega_n)^l}{\sqrt{1-\zeta^2}} \sin\left(\overline{l-1}\,\theta\right), l = 1,2,\ldots,$$

where

$$\theta = \tan^{-1} \frac{\sqrt{1-\zeta^2}}{\zeta}.$$

This sequence is oscillatory with period, $2\pi/\theta$, which changes from 4 to ∞ as ζ changes from 0 to 1. The effect of negative λ is to reduce the term $\frac{\sqrt{1-\zeta^2}}{\zeta}$, and theoretically as $\lambda \to -\infty$, $\theta \to 0$, and the Markov-Poisson parameter sequence (MPPS) becomes non-oscillatory. A large β_c can then make the sequence fast converging (Figure 2.1b).

Remark 2.1

As seen from Figure 2.1, β_c (> 1) compresses the s-plane so that the ZOC is expanded to a new circle of radius β_c. Dhawan et al. (1991) appropriately called β_c a "band-compression factor."

Remark 2.2

Given that the Poisson filters are unstable for $\lambda < 0$, their outputs may grow drastically, amplifying numerical errors. As is done in the case of integral-equation approaches (Mukhopadhyay and Rao 1991), this may be controlled by periodically resetting of the filter outputs to zero. But this requires estimation of initial conditions.

However, one has to rely on the prior knowledge of the system poles to choose β_c and λ. In the sequel, "Markov" parameters mean "Markov-Poisson" parameters only.

For the ith MISO subproblem, from Eq. (2.7), considering d_{ij} ($\geq 2n_{ij}$) MPs of the jth subsystem, and dropping the subscript "i" in all relevant symbols, we have the identification model

$$\tilde{y}(t) = \sum_{j=1}^{v_i} \sum_{l=1}^{d_j} h_{l,j} u_j^l(t) = \Phi^T(t)\theta,$$

where

$$\Phi(t) = [\Phi_1(t), \ldots, \Phi_{v_i}(t)]^T,$$

$$\Phi_j(t) = [\mathcal{F}_1(p)u_j(t), \ldots, \mathcal{F}_{d_j}(p)u_j(t)]^T, \; j = 1, \ldots, v_i,$$

$$\theta = \left[h_{1,1} \cdots h_{d_1,1} \middle| \cdots \middle| h_{1,v_i} \cdots h_{d_{v_i},v_i}\right]^T,$$

$$\mathcal{F}_l(p) = \left(\frac{\beta_c}{p+\lambda}\right)^l$$

For $i = 1,2,\ldots,v_o$, a set of independent estimation equations arises, which can be solved individually.

Remark 2.3

Note the decoupling of the identification model into noninteracting SISO elements. This is due to the chosen linear parameterizations where the structure has no unknown denominator.

2.2.3 Markov Parameter Estimation

Discretizing the differentiation operator p and time t via BPFs (Rao 1983) where

$$p^{-1}f(t) = \frac{T_s}{2}\left(\frac{1+q^{-1}}{1-q^{-1}}\right)f(k)$$

and where $f(k)$ denotes the block pulse value of $f(t)$ at $t = kT_s$, over the interval $\left[\overline{k-1}\,T_s, kT_s\right)$, with T_s denoting the block-pulse width and q^{-1} denoting the usual backward shift operator,

$$\mathcal{F}_l\left(q^{-1}\right) = \left[\frac{\beta_c T_s\left(1+q^{-1}\right)}{(\lambda T_s + 2) + (\lambda T_s - 2)q^{-1}}\right]^l$$

and the identification model is

$$\tilde{y}(k) = \Phi(k)^{\mathrm{T}}\theta, \tag{2.10}$$

with $\mathcal{F}_l(p)$ replaced by $\mathcal{F}_l\left(q^{-1}\right)$.

Referring to the above, define a cost function

$$J(\theta) = \left[\theta - \hat{\theta}(0)\right]^{\mathrm{T}}\mathbf{P}^{-1}(0)\left[\theta - \hat{\theta}(0)\right] + \sum_{k=1}^{N}\left[y(k) - \tilde{y}(k)\right]^2.$$

The LS estimate that minimizes $J(\theta)$ is

$$\hat{\theta}(N) = \left[\mathbf{P}^{-1}(0) + \sum_{k=1}^{N}\Phi(k)\Phi(k)^T\right]^{-1}\left[\mathbf{P}^{-1}(0)\hat{\theta}(0) + \sum_{k=1}^{N}\Phi(k)y(k)\right] \tag{2.11}$$

provided the inverse exists. The above estimate may be calculated using the conventional RLS algorithm. In the subsequent sections, we refer to this recursive algorithm as $\mathcal{A}\left\{u(k), y(k), \theta; k \in [1, N]\right\}$.

Remark 2.4 (Numerical behavior of the estimation algorithm)

This is dictated by the condition number of the matrix $\left[\mathbf{P}^{-1}(0) + \sum_{k=1}^{N}\Phi(k)\Phi(k)^{\mathrm{T}}\right]$. *Use of BPF as the basis is observed to result in significantly large condition numbers of this matrix (Subrahmanyam and Rao 1993) given that this basis is overlapping and non-orthogonal. Square root [such as Polter's (Bierman 1977)] implementation of the RLS algorithm is recommended to avoid possible ill conditioning. See Section 3.5 for a discussion of these numerical issues.*

2.2.4 Identification of Structure

The first step toward parametric identification is identifying the structure of the realization either in state–space or as a frequency response function (i.e., TFM). For SISO systems, this step degenerates to the determination of the system order that is the only structural invariant. In the multivariable case, a set of indices that evolve from the specified canonical form is to be determined. For the controller canonical form (see the following section) realization, this is equivalent to finding the controllability indices $v_j, j = 1,2,\ldots,m$, which are the orders of the column-wise CDs of the TFM (Wolovich 1974).

One of the obvious ways of determining these controllability indices is to examine the columns of the Hankel matrix $\mathcal{H}(p, q)$ formed from the MPS as

$$\mathcal{H}(p, q) = \begin{bmatrix} \mathbf{H}_1 & \mathbf{H}_2 & \cdots & \mathbf{H}_q \\ \mathbf{H}_2 & \mathbf{H}_3 & \cdots & \mathbf{H}_{q+1} \\ \vdots & \vdots & \cdots & \vdots \\ \mathbf{H}_p & \mathbf{H}_{p+1} & \cdots & \mathbf{H}_{p+q} \end{bmatrix} \tag{2.12}$$

for predecessor independence. In view of the decomposition into MISO models,

$$\mathbf{H}_l = \left[h_{l,1}, h_{l,2}, \ldots h_{l,m}\right].$$

Interchanging columns, Eq. (2.12) may be written as

$$\mathcal{H}(p,q) = \left[\mathcal{H}_1, \mathcal{H}_2, \ldots, \mathcal{H}_{v_i}\right],$$

where $\mathcal{H}_j$, $j = 1,2,\ldots,v_i$ are the $(p \times q)$ Hankel matrices of row subsystems of the MISO submodel.

Thus, the problem of the structural identification of the MISO system is also decomposed into equivalent problems of finding ranks of Hankel matrices of individual row subsystems. Singular value decomposition may be used for this purpose.

According to the "partial realization" theory introduced by Tether (1970) and Kalman (1970), given a finite sequence of MPs it is possible to find a finite dimensional realization $\{\mathbf{A}, \mathbf{B}, \mathbf{C}\}$ whose first few MPs are correspondingly equal to the given finite sequence of MPs. Accordingly, given a finite estimated MPS, irreducible TFM models can be derived using Eqs. (2.4) and (2.5). However, because the proposed MP model involves unmodeled dynamics, such a realization will be biased. To reduce this bias, a technique based on pole-placement is derived in the next section.

2.3 Finitization of the Markov Parameter Sequence

By definition, the MPS is of infinite length, and it is finite only when all the poles of each subsystem of the TFM lie at the origin of the convergence zone. Keeping this in mind, a method can be devised to finitize the sequence to the minimum possible value ν, given by

$$\nu = \begin{cases} \max_{i,j}\{n_{i,j}\} & \text{—when no common denominator (CD) is assumed} \\ \max_{j}\{n_{j}\} & \text{—when column-wise CD is assumed} \end{cases}$$

Consider the controller form realization {**A**, **B**, **C**} of the TFM $\mathbf{G}^0(s)$, where s is λ-shifted.

2.3.1 Controller Form Realization

The right matrix fraction description of $\mathbf{G}^0(s)$ is

$$\mathbf{G}^0(s) = \mathbf{N}(s)\mathbf{D}^{-1}(s)$$

where **D**(s) is both column and row reduced,

$$\mathbf{D}(s) = \text{blockdiag}\left[\mathbf{A}_j^*(s), j = 1,2,\ldots,\nu_i\right]$$

$\mathbf{A}_j^*(s)$ is the CD of the jth column of $\mathbf{G}^0(s)$, and

$$\mathbf{N}(s) = \left\{B_{ij}^*(s); B_{ij}^*(s) = \frac{B_{ij}(s)}{A_{ij}(s)}A_j^*(s), i = 1,2,\ldots,\nu_o; j = 1,2,\ldots,\nu_i\right\},$$

where

$$\mathbf{A}_j^*(s) = s^{n_j} + a_{1,j}^* s^{n_j - 1} + \cdots + a_{n_j,j}^*, \text{ and}$$

$$\mathbf{B}_{ij}^*(s) = b_{1,ij}^* s^{n_{ij} - 1} + \cdots + b_{n_{ij},j}^*.$$

It can be shown that the controller form realization of $\mathbf{G}^0(s)$ denoted by the triple {**A**, **B**, **C**} is

$$\mathbf{A}(s) = \text{blockdiag}\left[\mathbf{A}_j(s),(n_j \times n_j): j = 1,2,\ldots,\nu_i\right],$$

$$\mathbf{B}^{\mathrm{T}} = \text{blockdiag}\left\{[0,\ldots,0,1],(1 \times n_j), j = 1,2,\ldots,\nu_i\right\}, \text{ and}$$

$$\mathbf{C} = \{\mathbf{c}_{ij}, i = 1,2,\ldots,\nu_o, j = 1,2,\ldots,\nu_i\},$$

where

$$\mathbf{A}_j = \begin{bmatrix} \mathbf{0}_{n_j-1} & \mathbf{I}_{n_j-1} \\ & \mathbf{k}_j \end{bmatrix},$$

$$\mathbf{k}_j = \left[a^*_{n_j,j}, a^*_{n_j-1,j}, \ldots, a^*_{1,j}\right], \text{ and}$$

$$\mathbf{c}_{ij} = \left[b^*_{n_{ij},ij}, \ldots, b^*_{1,ij}, \ldots, 0, \ldots, 0\right].$$

For proof see Wolovich (1974, Section 3.4, 114).

The matrix $\mathbf{A}$ may be written as

$$\mathbf{A} = \mathbf{A}_0 - \mathbf{BK},$$

where

$$\mathbf{A}_0 = \text{blockdiag}\left\{\begin{bmatrix} 0 & 1 & 0 & \cdots & 0 \\ 0 & 0 & 1 & \cdots & 0 \\ \vdots & \vdots & \vdots & \cdots & \vdots \\ 0 & 0 & 0 & \cdots & 0 \end{bmatrix}, (n_j \times n_j), j = 1,2,\ldots,\nu_i\right\}, \text{ and}$$

$$\mathbf{K} = \left[\mathbf{k}_j, (1 \times n_j), j = 1,2,\ldots,\nu_i\right].$$

Therefore,

$$\dot{\mathbf{x}} = \mathbf{A}_0\mathbf{x}(t) - \mathbf{B}\bar{\mathbf{u}}(t), \tag{2.13}$$

where

$$\bar{\mathbf{u}}(t) = \mathbf{u}(t) - \mathbf{Kx}(t) \tag{2.14}$$

is the modified input signal. The solution of Eq. (2.13), neglecting initial conditions, is

$$\mathbf{x}(t) = \left[\bar{u}_1^{n_1}(t), \bar{u}_1^{n_1-1}(t), \ldots, \bar{u}_1^1(t), \middle| \cdots \middle| \bar{u}_{\nu_i}^{n_{\nu i}}(t), \bar{u}_{\nu_i}^{n_{\nu i}-1}(t), \ldots, \bar{u}_{\nu_i}^1(t)\right]. \tag{2.15}$$

The fictitious system described by the signal pair $\bar{\mathbf{u}}(t)$ and $\mathbf{y}(t)$ has a finite MPS, and by transforming the original system into the one described by Eq. (2.13), the approximation error in model (2.7) can be made to vanish. We now have the following iterative algorithm for the MISO case.

Algorithm

The proposed algorithm has the following steps:

1. Set iteration count $\xi = 1$. Initialize $\hat{\mathbf{k}}_{\xi-1} = [0,\ldots,0]$.
2. Estimate $\hat{\theta}_{\xi}$ using $A\{\mathbf{u}(k) - \hat{\mathbf{k}}_{\xi}\hat{\mathbf{x}}(k), \mathbf{y}(k), \boldsymbol{\theta}_{\xi}; k \in [N_{\xi-1}, N_{\xi}]\}$.
3. Form $\Delta\hat{\mathbf{k}}_{\xi} = [\hat{a}_{n_1,1}\cdots\hat{a}_{1,1}|\cdots|\hat{a}_{n_{\nu_i},\nu_i}\cdots\hat{a}_{1,\nu_i}]$.
   Where the right-hand side vector is calculated from $\hat{\theta}_{\xi}$, using Eqs. (2.4) and (2.5), and update $\hat{\mathbf{k}}_{\xi}$,
   $$\hat{\mathbf{k}}_{\xi} = \hat{\mathbf{k}}_{\xi-1} + \Delta\hat{\mathbf{k}}_{\xi}.$$
4. Check if MPS is finitized. If yes, go to Step 5; else, set $\xi \to \xi + 1$ and go to Step 2.
5. Stop.

Remark 2.5

The above algorithm clearly uses the pole-placement technique for iteratively shifting the poles to the center of the ZOC. While doing so, the bias of the estimated model in each iteration is reparameterized to be estimated in the next iteration. To illustrate this idea of bias reparameterization in this algorithm, consider, for example, the SISO case,

$$y(s) = \frac{B(s)}{A(s)}u(s). \tag{2.16}$$

Let $\frac{\hat{B}_1(s)}{\hat{A}_1(s)}$ *be an initial estimate of* $\frac{B(s)}{A(s)}$*, which may be obtained from a truncated MP model and*

$$\Delta_1(s) = \frac{B(s)}{A(s)} - \frac{\hat{B}_1(s)}{\hat{A}_1(s)} \tag{2.17}$$

be a measure of bias/error. Then

$$y(s) = \frac{\hat{B}_1(s) + \hat{A}_1(s)\Delta_1(s)}{s^n}\bar{u}(s), \tag{2.18}$$

where

$$\bar{u}(s) = \frac{s^n}{\hat{A}_1(s)}u(s). \tag{2.19}$$

Note that the reference system is changed from $\frac{B(s)}{A(s)}$ *to* $\frac{\hat{B}_1(s)+\hat{A}_1(s)\Delta_1(s)}{s^n}$*, which is a function of the unknown error* $\Delta_1(s)$ *and the known estimate* $\hat{B}_1(s)$ *and* $\hat{A}_1(s)$*. When* $\Delta_1(s)$ *becomes zero* $\forall s$*, then and only then the reference system has a finite MPS.*

Remark 2.6

In Step 2, the state vector $\hat{\mathbf{x}}(t)$ *is approximated as in Eq. (2.15). Given that the initial conditions of state and those of Poisson filters are neglected in doing so, the initial state model is to be estimated. For this, the data vector* $\mathbf{\Phi}(t)$ *and the parameter vector* θ *are augmented by* $\mathcal{I}(t) = [\mathcal{I}_1(t), \ldots, \mathcal{I}_{d_c}(t)]^T$ *and* $[c_1(t), \ldots, c_{d_c}(t)]^T$ *respectively, where*

$$\mathcal{I}_i(t) = e^{-\lambda t} \beta_c^i \frac{t^{i-1}}{(i-1)!}$$

and $c_1(t), \ldots, c_{d_c}(t)$ *are the parameters corresponding to the initial conditions. It is observed that convergence of the estimator improves by increasing* d_c*.*

Remark 2.7

In Step 4, the covariance matrix $\mathbf{P}$ *is reset according to the following law:*

$$\mathbf{P} = \alpha \rho^{\xi-1} \mathbf{I},$$

where α *is large and* $0 \le \rho \le 1$*.*

Remark 2.8

At the end of the run, $\hat{\theta}_\xi$ *converges to* $[\hat{b}_{1,1}, \ldots, \hat{b}_{n_1,1}, 0, \ldots, 0 | \cdots | \hat{b}_{1,\nu_i}, \ldots, \hat{b}_{n_{\nu_i},\nu_i}, 0, \ldots 0]$ *and* $\hat{\mathbf{k}}_\xi$ *converges to* $[\hat{a}_{n_1,1}, \ldots, \hat{a}_{1,1} | \cdots | \hat{a}_{n_{\nu_i},\nu_i}, \ldots, \hat{a}_{1,\nu_i}]$*. This suggests a suitable stopping test of this algorithm.*

2.4 Identifiability Conditions

$\hat{\theta}$ given by Eq. (2.11) is identifiable if the inverse on the right-hand side exists. A sufficient condition is

$$\mathbf{R} = \sum_{k=1}^{N} \Phi(k)\Phi(k)^T > 0.$$

We now derive conditions on the input such that the above exists.

Lemma 2.1

For $j = 1,2,\ldots,\nu_i$,

$$\boldsymbol{\Phi}_j(k) = \frac{1}{D_j(q^{-1})}\mathbf{F}_{d_j}\begin{bmatrix} u_j(k) \\ u_j(k-1) \\ \vdots \\ u_j(k-d_j) \end{bmatrix}, \tag{2.20}$$

where

$$D_j(q^{-1}) = \left[(\lambda T_s + 2) + (\lambda T_s - 2)q^{-1}\right]^{d_j}, \tag{2.21}$$

$$\mathbf{F}_{d_j} = \begin{bmatrix} f_{0,1} & f_{1,1} & \cdots & f_{d_j,1} \\ \vdots & \vdots & \vdots & \vdots \\ f_{0,d_j} & f_{0,d_j} & \cdots & f_{d_j,d_j} \end{bmatrix}_{d_j \times (d_j+1)}, \tag{2.22}$$

and

$$f_{h,i} = \left(\sum_{r=\nu}^{\min(h,i)} {}^{i}C_r \, {}^{d-i}C_{h-r} \left(\lambda T_s - 2\right)^{h-r} \left(\lambda T_s + 2\right)^{d-i-h+r}\right)\beta_c^i T_s^i, \tag{2.23}$$

where

$$\nu = \begin{cases} 0 & \text{if } h \le d-i \\ h-d+i & \text{otherwise} \end{cases}.$$

Proof

For $1 \le i \le d_j$,

$$\mathcal{F}_i\left(q^{-1}\right) = \left[\frac{\left[\beta_c T_s(1+q^{-1})\right]^i \left[(\lambda T_s + 2) + (\lambda T_s - 2)q^{-1}\right]^{d_j - i}}{D_j(q^{-1})}\right].$$

$$= \frac{\sum_{h=0}^{d_j} f_{h,i} q^{-1}}{D_j(q^{-1})}$$

Therefore,

$$\mathcal{F}_i\left(q^{-1}\right)u_j(k)=\frac{1}{D_j(q^{-1})}\left[f_{0,i},f_{1,i},\ldots,f_{d_j,i}\right]\begin{bmatrix}u_j(k)\\u_j(k-1)\\u_j(k-d_j)\end{bmatrix}. \tag{2.24}$$

Arranging $\mathcal{F}_i\left(q^{-1}\right)u_j(k)$ row-wise for $1\le i\le d_j$, we obtain Eq. (2.20).

QED

Lemma 2.2

$\rho\left(\mathbf{F}_{d_j}\right)=d_j$ where $\rho(\bullet)$ denotes rank.

Proof

By induction, noting that, for some h,

$$\mathbf{F}_{h+1}=\begin{bmatrix} & \mathbf{F}_h & & \begin{matrix}0\\0\\\vdots\end{matrix}\\ 1 & f_1 & f_2 & f_{h+1}\end{bmatrix}\begin{bmatrix}(\lambda T_s+2) & (\lambda T_s-2) & \cdots & 0\\ \vdots & (\lambda T_s+2) & \cdots & 0\\ \vdots & \vdots & \cdots & \vdots\\ 0 & 0 & \cdots & (\lambda T_s-2)\\ 0 & 0 & \cdots & (\lambda T_s+2)\end{bmatrix}_{h+2},$$

where

$$f_i=\sum_{r=0}^{i}{}^{h+1}C_r(-1)^{i-r}\frac{(\lambda T_s-2)^{i-r}}{(\lambda T_s+2)^{i-r+1}}.$$

QED

Lemma 2.3

Let $u_j(k)$ be persistently exciting of order m. Assume that $H\left(q^{-1}\right)$ is an asymptotically stable filter with no zeros on the unit circle. Then the filtered signal $u_j'(k)=H\left(q^{-1}\right)u_j(k)$ is also persistently exciting of order m.

Proof

See Söderström and Stoica (1989, 123).

QED

Now,

$$\mathbf{R} = N\,\bar{\varepsilon}\left\{\Phi(k)\Phi(k)^{\mathrm{T}}\right\}$$

$$= N\,\bar{\varepsilon}\left\{\begin{bmatrix} \Phi_1(k)\Phi_1(k)^{\mathrm{T}} & \cdots & \Phi_1(k)\Phi_{v_i}(k)^{\mathrm{T}} \\ \vdots & \cdots & \vdots \\ \Phi_{v_i}(k)\Phi_1(k)^{\mathrm{T}} & \cdots & \Phi_{v_i}(k)\Phi_{v_i}(k)^{\mathrm{T}} \end{bmatrix}\right\},$$

where

$$\bar{\varepsilon}\left\{\Phi_i(k)\Phi_j(k)^{\mathrm{T}}\right\} = \frac{1}{D_i(q^{-1})}\mathbf{F}_{d_i}\begin{bmatrix} r_{ij}(0) & \cdots & r_{ij}(d_i) \\ \vdots & \cdots & \vdots \\ r_{ij}(-d_j) & \cdots & r_{ij}(d_i - d_j) \end{bmatrix}\mathbf{F}_{d_j}^{\mathrm{T}}\frac{1}{D_j(q^{-1})},$$

$$r_{ij} = \bar{\varepsilon}\left\{u_i(k)u_j(k+\tau)\right\},$$

and $\bar{\varepsilon}$ is defined as

$$\bar{\varepsilon}\left\{x(k)\right\} = \frac{1}{N}\sum_{k=1}^{N} x(k).$$

Now we have the following theorem.

Theorem 2.1

(Sufficiency) Under the following conditions, **R** is *nonsingular*

1. u_j *is persistently exciting of order* $(d_j + 1)$.
2. $r_{ij}(\tau) = 0, \forall j,$ where $0 \le \tau \le \max(d_i, d_j)$.

Remark 2.9

The above conditions imply that the spectrum of $u_i(k)$ *have* $\max(d_j) + 1$ *points of support that are non-overlapping with those of* $u_j(k), (j \ne i)$.

Remark 2.10

It can be verified that $\bar{u}_j(k)$ *has the same order of persistency as that of* $u_j(k)$ *in view of Lemma 2.3, and thus the result of Theorem 2.1 remains unchanged for* $\xi > 1$.

2.5 Convergence Analysis of the Algorithm

The key idea behind the bias reparameterization algorithm is explained in Remark 2.5. To make the analysis simple, assume that (a) the model set is valid and (b) the input signal is sufficiently rich such that the estimator $\mathcal{A}$ converges.

Consider the SISO case with a $2n$-dimensional MP model and follow the notation in Eqs. (2.16)–(2.19).

Let

$$\mathcal{C}^n = \begin{Bmatrix} A_i(s) \mid A_i(s) = s^n + a_{1,i}s^{n-1} + a_{2,i}s^{n-2} + \cdots + a_{n,i} \\ s \in \mathcal{C}; a_i \in \mathbb{R} \end{Bmatrix}$$

and define a metric d on $\mathcal{C}^n$ as

$$d\left(A_i, A_j\right) = \left[\sum_{l=1}^{n} \left(a_{l,i} - a_{l,j}\right)\right]^{\frac{1}{2}} \quad \forall i, j. \tag{2.25}$$

Let $\mathcal{D}(A_0;\eta) = \{A \in \mathcal{C}^n \mid d(A, A_0) \le \eta\}$ be a closed ball, where A_0 is the center and η the radius. Also let $G^0(s) = \frac{B^0(s)}{A^0(s)}$ be the true system, where $A_0(s) \in \mathcal{D}$ and $\eta = d(A_0, 0)$.

Lemma 2.4

(a) C^n is a complete metric space with the metric defined in Eq. (2.25), and (b) the subspace $\mathcal{D}$ of the complete metric space C^n is a complete metric space.

Proof

1. d satisfies all the axioms of a metric, namely, for $A_i, A_j, A_l \in C^n$; d is real-valued, finite and non-negative; $d\left(A_i, A_j\right) = 0 \Rightarrow A_i = A_j$; $d\left(A_i, A_j\right) = d\left(A_j, A_i\right)$; and $d\left(A_i, A_j\right) \le d\left(A_i, A_l\right) + d\left(A_l, A_j\right)$.
2. Completeness of C^n: Consider any Cauchy sequence $\{A_i\}$ in C^n. Then for every $\varepsilon > 0$, there is an N such that, for $i, j < N$,

$$d\left(A_i, A_j\right) = \left[\sum_{l=1}^{n} \left(a_{l,i} - a_{l,j}\right)^2\right]^{\frac{1}{2}} < \varepsilon \tag{2.26}$$

or for $i, j > N$ and $l = 1, 2, \ldots, n$,

$$\left|a_{l,i} - a_{l,j}\right| < \varepsilon.$$

This shows that for every fixed $l,(1 \le l \le n)$, the sequence $a_{l,1}, a_{l,2} \cdots$ is a Cauchy sequence in $\mathbb{R}$. Given that $\mathbb{R}$ is a complete metric space, say $a_{l,i} \to a_i$ as $i \to \infty$. Using these n limits, define

$$A(s) = 1 + a_1 s + \cdots + a_n s^n.$$

Clearly $A \in \mathcal{C}^n$. Therefore, with $i \to \infty$, $d(A_j, A) \le \varepsilon, j > N$. This shows that A is the limit of $\{A_j\}$ and thus proves the completeness of $\mathcal{C}^n$.

QED

Let $\mathcal{T}$ be a mapping $\mathcal{T} : \mathcal{D} \to \mathcal{D}$ consisting of Steps 2 and 3 of the proposed algorithm. Given an initial guess that $A_1 \in \mathcal{D}$, of A_0, $\mathcal{T}$ produces the following iterative sequence,

$$A_1, A_2 = \mathcal{T} A_1, A_3 = \mathcal{T} A_2 = \mathcal{T}^2 A_1, \ldots$$

It is required to show that this sequence is Cauchy. It will then be argued that this sequence converges to the fixed point of $\mathcal{D}$ under the following conditions:

Condition 2.1

For any $A_i \in \mathcal{D}$, $d(A_i, A^0) \le d(A^0, 0)$, where $0 \in \mathcal{D}$ is the null element.

Condition 2.2

The order of the system $G^0(s)$ is known and is equal to the order of the model.
Given the MP series expansion of $G^0(s)$, $G^0(s) = \sum_{l=1}^{\infty} h_l s^{-l}$, define

$$r_0 = \left[\sum_{l=2n+1}^{\infty} h_l^2 \right]^{\frac{1}{2}}.$$

Also, for a given $A \in \mathcal{D}$, consider $G^0(s)\frac{A_i(s)}{s^n} = \sum_{l=1}^{\infty} h_{l,i} s^{-l}$, and define

$$r(A_i) = \left[\sum_{l=2n+1}^{\infty} h_{l,i}^2 \right]^{\frac{1}{2}}.$$

Remark 2.11

In some qualitative sense, $r(A_i)$ indicates the amount of undermodeling in the i-th iteration.

Lemma 2.5

A^0 is the unique fixed point of $\mathcal{D}$ under the mapping $\mathcal{T}:\mathcal{D}\to\mathcal{D}$.

Proof

A^0 in the denominator of $G^0(s)$ can be cancelled by $A^0(s)$ alone, which is possible under Condition 2.2. Hence, $\mathcal{T}\,A^0 = A^0 \Rightarrow A^0$ is the unique fixed point.

QED

Lemma 2.6

For any $A_i \in \mathcal{D}$, $r(A_i) < r_0$, under Condition 2.1.

Proof

Without loss of generality, let $B^0(s) = 1$. Consider

$$\frac{1}{A^0(s)}\frac{A_i(s)}{s^n} = \frac{s^n + a_{1,i}s^{n-1} + \cdots + a_{n,i}}{s^n\left(s + a_1 s^{n-1} + \cdots + a_n\right)} = \sum_{l=1}^{\infty} h_{l,i}s^{-l}.$$

Then,

$$h_{l,i} = \begin{cases} 0 & l = 1,\ldots,n-1 \\ 1 & l = n \\ \left(a_{l-n,i} - a_{l-n}\right) - \sum_{k=1}^{l-n-1} h_{n+k,i}a_{l-n-k} & l = n+1,\ldots,2n \\ -\sum_{k=1}^{l-n-1} h_{n+k,i}a_{l-n-k} & l = 2n+1,\ldots \end{cases} \quad (2.27)$$

Let

$$\frac{1}{s^n A^0(s)} = \sum_{l=1}^{\infty} h_l s^l.$$

Then,

$$h_{l,i} = \begin{cases} 0 & l = 1,\ldots,n-1 \\ 1 & l = n \\ a_{l-n} - \sum_{k=1}^{l-n-1} h_{n+k,i} a_{l-n-k} & l = n+1,\ldots,2n \cdot \\ -\sum_{k=1}^{l-n-1} h_{n+k,i} a_{l-n-k} & l = 2n+1,\ldots \end{cases} \tag{2.28}$$

Comparing Eqs. (2.27) and (2.28), under Condition 2.1,

$$h_{l,i} \le h_l, \quad i = n+1,\ldots$$

$$\Rightarrow r(A_i) < r_0.$$

QED

Lemma 2.7

Given any $A_i \in \mathcal{D}$, *let* $A_{i+1} = \mathcal{T}\ A_i$. *Then*

$$d(A_{i+1}, A^0) = \alpha\ d(A_i, A^0) \tag{2.29}$$

where $\alpha < 1$.

Proof

The mapping $\mathcal{T}$ estimates $\frac{A_i(s)}{s^n A^0(s)}$ as $\frac{1}{s^n + A_{i+1}(s) - A_i(s)}$, or in other words, $A^0(s)$ is estimated as $A_{i+1}(s) = A_i(s) + \Delta A_i(s)$, where

$$\Delta A_i(s) = \frac{A^0(s) - A_i(s)}{A_i(s)} s^n.$$

Given that $r(A_i) < r_0$ (see Lemma 2.6),

$$d(A_{i+1}, A^0) = d(\mathcal{T}\ A_i, A^0) \le d(A_i, A^0)$$

or with some $\alpha < 1$,

$$d(A_{i+1}, A^0) = \alpha d(A_i, A^0).$$

QED

Remark 2.12

In the proposed algorithm, the mapping $\mathcal{T}$ allows $A^0(s)$ to be estimated at reduced bias, each time, because of reduced undermodeling (see Lemma 2.6). As a result, as each time $\mathcal{T}$ is applied, the estimate moves toward A^0 (in the sense of distance). This justifies the result of the above lemma.

We now have the following theorem:

Theorem 2.2

(Sufficiency) Choose any $A_i \in \mathcal{D}$ and construct the iterative sequence $\{A_i\}$,

$$A_1, A_2 = \mathcal{T} A_1, A_3 = \mathcal{T} A_2 = \mathcal{T}^2 A_1, \ldots \tag{2.30}$$

Then $\{A_i\}$ converges to the fixed point A^0 under Conditions 2.1 and 2.2.

Proof

From Eq. (2.29), following the usual way (e.g., see Kreyszig 1978), A_i may be shown to be Cauchy.

Because $\mathcal{D}$ is a complete subspace of C^n, $\{A_i\}$ is convergent, and because there is a unique fixed point A^0, converges to A^0.

QED

2.6 Illustrative Examples

Example 2.1 (Iterative bias reparameterization):

Consider the following CT SISO system:

$$G^0(s) = \frac{2}{s^2 + 3s + 2}.$$

This system is simulated using a pseudo random binary sequence (PRBS) at the sampling rate $T_s = 0.01$ s. The proposed bias reparameterization algorithm is used to estimate the first four MPs of $G^0(s)$ with $\lambda = 1.0$ and $\beta_c = 4.0$ for five iterations, for 1,000 recursions per iteration. Table 2.1 shows the estimates in their rational transfer function format for four iterations, and the pattern of parameter convergence is shown in Figure 2.2. In this figure, each 10 s on the time axis corresponds to one iteration of the algorithm. The effect of bias reparameterization for finitization of MPS is clearly noticeable in this figure, as $\bar{h}_2, \bar{h}_3$, and $\bar{h}_4$ approach zero with iterations.

TABLE 2.1

Parameter Estimates (Example 2.1)

Iteration no.	$b_1(0.0)$	$b_2(2.0)$	$a_1(3.0)$	$a_2(2.0)$
1	0.0359	1.8318	2.6240	1.8088
2	0.0115	1.9087	2.8049	1.8785
3	0.0039	1.9616	2.9043	1.9345
4	0.0434	1.9686	2.9490	1.9635
6	0.0029	1.9733	2.9649	1.9682

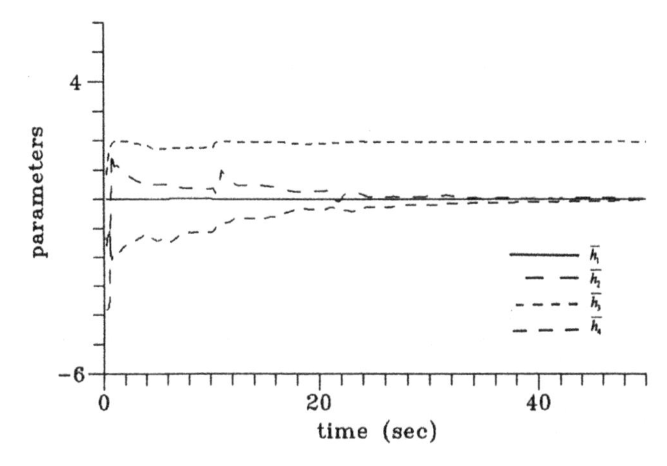

FIGURE 2.2
Pattern of MP convergence (Example 2.1).

Example 2.2 (Structure identification and parameter estimation):

Consider

$$G^0(s) = \left[\frac{s^2+s+1}{s^3+4s^2+5s+2} \quad \frac{1}{s^2+2.5s+1.5} \right].$$

This system is simulated using PRBS inputs having low-mutual correlation at a sampling rate of 0.02 s. Poisson filter parameters are chosen as $\lambda = 1.0$ and $\beta_c = 3.0$. To begin with, the orders of the subsystems are found by estimating $d_1 = 7$ and $d_2 = 5$ MPs. The singular values of the associated Hankel matrix are found to be {8.97, 2.77, 0.32, 0.05} and {1.18, 0.87, 0.004}, respectively, for both subsystems. Considering this, a choice of $n_1 = 3$ and $n_1 = 2$ is made.

The parameter estimation experiment is then conducted with $d_1 = 6$, $d_2 = 4$, and $d_c = 4$, resulting in 14 parameters. The proposed algorithm

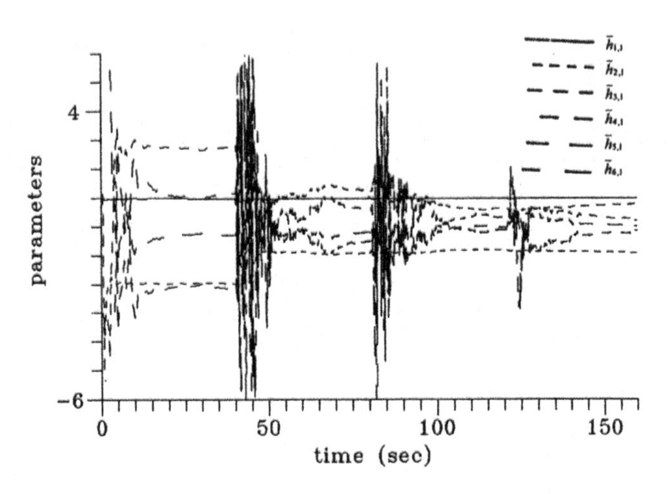

FIGURE 2.3
Pattern of MP convergence—Subsystem 1 (Example 2.2).

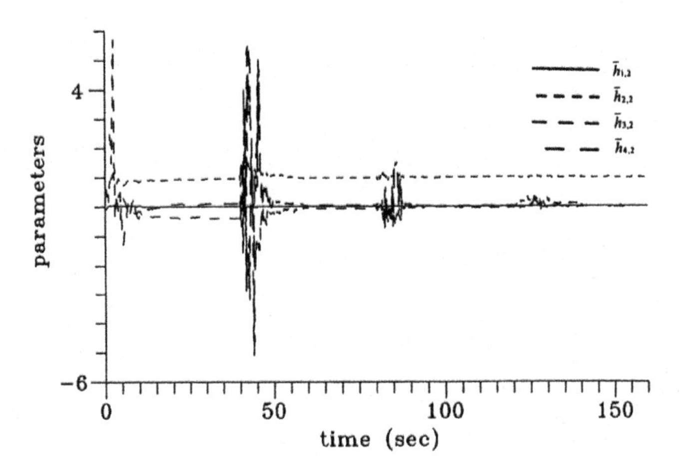

FIGURE 2.4
Pattern of MP convergence—Subsystem 2 (Example 2.2).

for the finitization of MPSs is applied for 160 s for four iterations, that is, each iteration is made on 40-s data. Figures 2.3 and 2.4 show the iteratively finitized MPS for Subsystem 1 and Subsystem 2, respectively, at 10% Gaussian white measurement noise. Note that the sequence is finitized to a sequence of 3 + 2 = 5 parameters, improving the quality of estimates. The corresponding transfer function parameter estimates are shown in Tables 2.2 and 2.3 for Subsystems 1 and 2, respectively.

TABLE 2.2

Parameter Estimates—Subsystem 1 (Example 2.2)

Iteration no.	$b_{1,1}(1.0)$	$b_{2,1}(1.0)$	$b_{3,1}(1.0)$	$a_{1,1}(4.0)$	$a_{2,1}(5.0)$	$a_{3,1}(2.0)$
1	0.9945	1.0895	1.1448	4.0712	5.5372	2.2335
2	0.9841	1.1448	0.8298	4.1740	5.2440	1.9528
3	0.9834	0.9878	0.9409	4.0080	5.0938	1.8866
4	0.9799	1.0117	0.9954	3.9655	5.1403	1.9182

TABLE 2.3

Parameter Estimates—Subsystem 2 (Example 2.2)

Iteration no.	$b_{1,2}(0.0)$	$b_{2,1}(1.0)$	$a_{1,2}(2.5)$	$a_{2,2}(1.5)$
1	0.0065	0.9665	2.4133	1.4791
2	−0.0003	0.9871	2.4797	1.4856
3	−0.0016	0.9996	2.5132	1.4946
4	−0.0012	0.9923	2.5116	1.4917

Example 2.3 (Quality of estimation)

To check the consistency and to compare the quality of estimates, we study the following example system taken from Mukhopadhyay et al. (1991):

$$G^0(s) = \begin{bmatrix} \frac{1}{s^2+3s+2} & \frac{1}{s+2} \end{bmatrix}.$$

Monte Carlo simulations of 20 runs each are made at different noise levels and the results are shown in Table 2.4. Here $\lambda = 1.0$, $\beta_c = 4.0$, $d_1 = 4, d_2 = 2$, and $d_c = 2$ are chosen. Data is obtained at $T_s = 0.01$ s. Each run is made

TABLE 2.4

Results of Monte Carlo Experiments with MP Approach (Example 2.3)

Noise level (%)	$b_{1,1}(0.0)$	$b_{2,1}(1.0)$	$a_{1,1}(3.0)$	$a_{2,1}(2.0)$	$b_{1,2}(1.0)$	$a_{1,2}(2.0)$
0	−0.0009	0.9956	2.9288	1.9485	0.9959	2.0002
10	0.0008	1.0008	2.9309	1.9504	0.9955	1.9985
20	0.0027	1.0051	2.9288	1.9559	0.9951	1.9968
50	0.0095	1.0103	2.8916	1.9944	0.9942	1.9915
100	0.0252	0.9952	2.7548	2.1840	0.9934	1.9797

TABLE 2.5

Comparison of Percentage REN (Example 2.3)

Submodel no.	Noise	DU Approach			MP
		LS/LS	GLS/LS	GLS/GLS	
1	0	0.35	0.32	0.34	2.34
	10	30.90	2.13	4.52	2.27
	20	62.83	24.64	24.64	2.24
	50	86.80	35.27	35.35	2.92
2	0	0.09	0.10	0.10	0.18
	10	20.29	2.92	9.15	0.21
	20	48.04	11.08	6.12	0.26
	50	75.85	46.44	29.34	0.46

for four iterations with 2,000 recursions per iteration. The results can be better appreciated by comparing these with those obtained under EE formulation. Table 2.5 shows the values of relative error norm (REN) defined as

$$REN = \frac{\left\| \theta_{ab} - \hat{\theta}_{ab} \right\|_2}{\left\| \theta_{ab} \right\|_2}$$

where $\theta_{ab} = \left[b_{1,1}, b_{2,1}, a_{1,1}, a_{2,1}, b_{1,2}, a_{1,2} \right]^T$, with LS and GLS schemes using the DU algorithm (taken from Mukhopadhyay et al. 1991), and the proposed MP-based approach. In this table, LS/LS specifies that the LS scheme is used in both stages of the DU algorithm. Similar notation applies for GLS/LS and GLS/GLS.

Example 2.4 (Resonant system):

Now consider a second-order resonant system (Wahlberg 1991), with $\zeta = 0.1$, $\omega_n = 1.0$, that is,

$$G^0(s) = \frac{1}{s^2 + 0.2s + 1}.$$

In this case, a choice of $\lambda = -2.0$, $\beta_c = 5.0$, and $d = 5$ is made. At the end of four iterations with 1,500 recursions per iteration over a data sampled at $T_s = 0.001$ s, the estimate is obtained as

$$\hat{G}(s) = \frac{-0.09581s + 1.07873}{s^2 + 0.1503s + 1.07879},$$

that is, $\hat{\zeta} = 0.09625 = 0.0962$ and $\hat{\omega}_n = 1.03864$.

2.7 Summary and Conclusions

A recursive algorithm for the parameter estimation of TFM in its irreducible form is presented in this chapter. It is clearly a dual of the pole placement problem in linear control design. This may be viewed as one of finding a suitable feedback gain for an "all-poles-at-the-origin" system to match the given input/output measurement data. This is solved backwards iteratively to derive the signal $\mathbf{u}(t)$ while updating $\hat{\mathbf{k}}$. As $\hat{\mathbf{k}}$ tends to $\mathbf{k}$, the signal pair $\{\bar{\mathbf{u}}(t), \mathbf{y}(t)\}$ tends to describe the dynamics of the all-poles-at-the-origin system. This is accomplished with a finitization of the estimated MPS as shown in Figures 2.2 through 2.4.

The results of Monte Carlo simulation experiments indicate the noise rejection properties of the proposed estimator. The REN is clearly the least in the case of MP models, as indicated in Table 2.5. Similar results in support of this were presented earlier by Subrahmanyam and Rao (1993), where comparisons were made with the PMF approach (Saha and Rao 1983) using both LS and IV techniques. The results are definitely comparable to those achieved with approaches like IV and BCLS (Sagara and Zhao 1989; Zhao et al. 1992).

Normally, MP models, like Laguerre models (Wahlberg 1991), are not suitable for identifying systems with poles (real/imaginary) very close to the imaginary axis because of the oscillatory nature of the sequence. However, it is shown in this chapter that, a suitable shift of the imaginary axis renders the estimation possible. Thus, a wider class of systems may be treated under the proposed identification framework.

3

Time Moment Models

3.1 Introduction

In this chapter, another new class of CTMs parameterized in terms of time moments of the system impulse response is proposed. Like Markov parameters, time moments also enjoy an important place in the field of reduced-order modeling (Jamshidi 1983, Bultheel and Van Barel 1986). Despite the wealth of other mathematically sound methods available for reduced order modeling, the moment matching method is still considered as the simplest and is widely used. However, these "not-directly-realizable" models had never attracted the attention of system identification researchers.

Most of the contents of this chapter run in parallel to those of the previous chapter. This is because of the structural similarities between the MP models and TM models. Compare Eqs. (1.13) and (1.15). Although MP models provide good high-frequency approximations, TM models, on the other hand, provide good low-frequency approximations.

The definition of time moments as in Eq. (1.16) appears to be attractive for their direct computation, given the impulse response. However, impulse response data is rarely available in practice. Moreover, TM models are not directly realizable as the basis is a chain of pure differentiators. To avoid this, a linear dynamic operation (Unbehauen and Rao 1987) is to be performed to obtain realizable models.

With this background, this chapter attempts to develop a useful approach for irreducible MIMO CT model identification. The material of this chapter is organized as follows:

- Section 3.2 begins with the basic definition of time moments of MIMO systems and their realization from state–space/TFM descriptions. To obtain realizable TM models, Poisson filtering is suggested.
- In Section 3.3, another version of the bias reparameterization algorithm is proposed for finitization of time-moment sequence (TMS). This algorithm iteratively cancels the denominators of the elements of the TFM to finitize the TMS.

- Results of numerical simulation experiments with the algorithm proposed above are presented in Section 3.4.
- The parameterizations presented in this and the previous chapters are compared in Section 3.5 for their predictive ability and their numerical behavior. Some guidelines on how to choose a basis of parameterization are also given.
- Finally, Section 3.6 presents some conclusions.

3.2 Time Moment Models

Considering Eq. (2.1), given the impulse response $\mathbf{g}^0(t)$, time moments evolve from the TFM as

$$\mathbf{G}^0(s) = \int_0^\infty \mathbf{g}^0(t)e^{-st}dt = \sum_{l=0}^{\infty} \mathbf{M}_l s^l, \tag{3.1}$$

where

$$\mathbf{M}_l = \frac{(-1)^l}{l!}\int_0^\infty t^l \mathbf{g}^0(t)dt. \tag{3.2}$$

Letting

$$A_{ij}(s) = 1 + a_{1,ij}s + \cdots + a_{n_{ij}ij}s^{n_{ij}}$$

$$B_{ij}(s) = b_{0,ij} + b_{1,ij}s + \cdots + b_{n_{ij}-1,ij}s^{n_{ij}-1}$$

and

$$\mathbf{M}_l = \left\{m_{l,ij}; i = 1,2,\cdots,\nu_o; j = 1,2,\cdots,\nu_i\right\},$$

we have

$$m_{l,ij} = b_{l,ij} - \sum_{r=0}^{l-1} m_{r,ij}a_{l-r,ij}; l = 1,2,\cdots,n_{ij-1} \tag{3.3}$$

and

$$m_{l+n_{ij},ij} = -\sum_{r=1}^{n_{ij}} m_{n_{ij}+l-r,ij}a_{r,ij}; l = 1,2,\cdots \tag{3.4}$$

Given the state–space description, we also have,

$$\mathbf{M}_l = \mathbf{C}\mathbf{A}^{-l+1}\mathbf{B},$$

In terms of $\{\mathbf{M}_l\}$, the system input–output relation is

$$\mathbf{y}(t) = \sum_{l=0}^{\infty} \mathbf{M}_l \mathbf{u}^{(l)}(t). \tag{3.5}$$

where $\mathbf{u}^{(l)}(t)$ is the lth time derivative of $\mathbf{u}(t)$. Assuming absolute convergence of TMS and thus uniform convergence of partial sums, we have the truncated TM model,

$$\tilde{\mathbf{y}}(t) = \sum_{l=0}^{d} \mathbf{M}_l \mathbf{u}^{(l)}(t) = \theta^{\mathrm{T}} \mathcal{B}(p)\mathbf{u}(t), \tag{3.6}$$

where $\theta = [\mathbf{M}_1, \mathbf{M}_2, \cdots, \mathbf{M}_d]^{\mathrm{T}}$ and $\mathcal{B} = \left[1, p, \cdots, p^d\right]^{\mathrm{T}}$.

To ensure validity of the above model even for systems with diverging TMS, additional exponential scaling of the series must be done to ensure its convergence. It can be verified that the TMS is absolutely convergent when $|\lambda_i(A)| > 1, \forall i$. To ensure convergence, "band expansion" is recommended. Recall Remark 2.1.

Remark 3.1

Further generalization of TM models by shifting the imaginary axis as is done in the case of MP models is, however, not possible. To verify this, consider

$$\mathbf{G}(s) = \sum_{l=0}^{\infty} \bar{\mathbf{M}}_l (s - \lambda)^l,$$

where

$$\bar{\mathbf{M}}_l = \sum_{l=0}^{\infty} {}^{i}C_{i-l} \lambda^{i-l} \mathbf{M}_i.$$

Given that the above summation runs from l to ∞, with truncated TM models, the desired shift cannot be accomplished. Compare with Eq. (2.8).

To avoid use of derivatives, Eq. (3.6) is operated on both sides by a $(d + 1)$th-order Poisson filter $\frac{\beta^{d+1}}{(s+\lambda)^{d+1}}$ (Saha and Rao, 1983). Here Poisson filtering is chosen only for its simplicity. The state variable filters (Young 1970, 1981) may be used for more flexible filtering. Denoting

$$\mathcal{F}_{l,d+1}(p) = \beta^{d+1}\frac{p^l}{(s+\lambda)^{d+1}}, \quad l = 0, 1, \cdots, d,$$

for the ith row of Eq. (3.6), and dropping the subscript "i" in all relevant symbols,

$$\mathcal{F}_{0,d+1}(p)\tilde{y}(t) = \sum_{l=0}^{d}\sum_{j=1}^{v_i} m_{l,j}\mathcal{F}_{l,d+1}(p)u_j(t).$$

In general, considering $d_j(\geq 2n_{ij})$ time moments of the jth subsystem and letting $d = \max_j\{d_j\}$,

$$\bar{y}^*(t) = \Phi^T\theta,$$

where

$$\bar{y}^*(t) = \mathcal{F}_{0,d+1}(p)\tilde{y}(t),$$

$$\Phi(t) = \left[\Phi_1(t), \cdots, \Phi_{v_i}(t)\right]^T,$$

$$\Phi_j(t) = \left[\mathcal{F}_{0,d+1}(p)u_j(t), \cdots, \mathcal{F}_{d_j,d+1}(p)u_j(t)\right]^T, j = 1, \ldots, v_i,$$

$$\theta = \left[m_{0,1}, \cdots, m_{d_1,1}, |\cdots|, m_{0,v_i}, \cdots, m_{d_{v_i},v_i}\right]^T.$$

Upon discretization, we have the identification model

$$\tilde{y}^*(k) = \Phi^T(k)\theta$$

with $\mathcal{F}_{l,d+1}(p)$ replaced by

$$\mathcal{F}_{l,d+1}\left(q^{-1}\right) = 2^l T_s^{d+1-l}\frac{\left(1-q^{-1}\right)^l\left(1+q^{-1}\right)^{d+1-l}}{\left[(\lambda T_s+2)+(\lambda T_s-2)q^{-1}\right]^{d+1}}.$$

Parameter estimation may then be carried out as

$$\hat{\theta}(N) = \left[\mathbf{P}(0)^{-1} + \sum_{k=1}^{N}\Phi(k)\Phi(k)^T\right]^{-1}\left[\mathbf{P}(0)^{-1}\hat{\theta}(0) + \sum_{k=1}^{N}\Phi(k)\mathcal{F}_{0,d+1}\left(q^{-1}\right)y(k)\right].$$

Remark 3.2 (Identifiability conditions)

$\hat{\theta}$ as given above exists provided u(k) satisfies the following conditions:

a. *u_j is persistently exciting of order $(d_j + 2)$.*
b. *$r_{ij}(\tau) = 0, \forall i \neq j$, where $0 \leq \tau \leq \max(d_i + 1, d_j + 1)$.*

Proof, being similar to the case of MP models, is omitted here.

3.3 Finitization of Time-Moment Sequence

We extend the idea of bias reparameterization of Section 2.3 to the case of time moments also. The TMS is finite when all the subsystems are denominator free; in which case, the length of the TMS is $\max_j\{n_j\}$ and modeling via Eq. (3.6) is exact. This situation can be met by adding fictitious zeros to each subsystem, to cancel their denominators. In an identification experiment, this is achievable for finite dimensional systems as illustrated below for the SISO case,

$$y(s) = \frac{B(s)}{A(s)} u(s).$$

If the denominator $A(s)$ is known, we can write

$$y(s) = \frac{B(s)}{A(s)} u(s) = \sum_{i=0}^{n-1} b_i s^i \bar{u}(s),$$

where $\bar{u}(s) = \frac{1}{A(s)} u(s)$. Thus, the model between $\bar{u}(t)$ and $y(s)$ has a finite TMS.

Let $\frac{\hat{B}_1(s)}{\hat{A}_1(s)}$ be the initial estimate of $\frac{B(s)}{A(s)}$, which may be obtained using a truncated model as described in the previous section. Then,

$$\begin{aligned}\frac{y(s)}{\bar{u}(s)} &= \frac{B(s)}{A(s)} \hat{A}_1(s) = \hat{A}_1(s)\Delta_1(s) + \hat{B}_1(s),\\ &= G_1(s) \text{ (say)}\end{aligned}$$

where

$$\Delta_1(s) = \frac{B(s)}{A(s)} - \frac{\hat{B}_1(s)}{\hat{A}_1(s)}$$

is a measure of the bias. Note that the reference system is changed from $\frac{B(s)}{A(s)}$ to $\{\hat{A}_1(s)\Delta_1(s)+\hat{B}_1(s)\}$ which is a function of the unknown bias $\Delta_1(s)$ and known estimate $\frac{\hat{B}_1(s)}{\hat{A}_1(s)}$.

When $\Delta_1(s)$ becomes zero ($\forall s$), then and only then the reference system becomes denominator free.

Let $\frac{\hat{B}_2(s)}{\Delta\hat{A}_1(s)}$ be the estimate of $G_1(s)$.

Then, we may update

$$\bar{u}(s)=\frac{1}{\Delta\hat{A}_1(s)}\frac{1}{\hat{A}_1(s)}u(s)=\frac{1}{\hat{A}_2(s)}u(s)$$

and the new estimate becomes $\frac{\hat{B}_2(s)}{\hat{A}_2(s)}$. This procedure may be repeated until the TMS is finitized.

We thus have the following algorithm.

Algorithm

1. Set iteration count $\xi=1$ and initialize $A_{\xi,j}(p)=1.0,\ j=1,2,\cdots,v_{\mathrm{i}}$
2. Estimate $\hat{\theta}_\xi$ using

$$\mathcal{A}\left\{\left[\frac{1}{A_{\xi,j}(\bar{p})}u_j(k),j=1,2,v_{\mathrm{i}}\right],y(k),\hat{\theta}_\xi;k\in\left[N_{\xi-1},N_\xi\right]\right\}$$

3. Calculate $\{\hat{a}_{i,j};i=1,2,\cdots,n_j;j=1,2,\cdots,v_{\mathrm{i}}\}$ from Eq. (3.4). Form $\Delta\hat{A}_{\xi,j}(\bar{p})=1+\hat{a}_{1,j}\bar{p}+\cdots+\hat{a}_{n_j,j}\bar{p}^{-n_j};j=1,2,\ldots,v_{\mathrm{i}}$.
4. Update $\hat{A}_{\xi+1,j}(\bar{p})=\hat{A}_{\xi,j}(\bar{p})\Delta\hat{A}_{\xi,j}(\bar{p});j=1,2,\ldots,v_{\mathrm{i}}$.
5. Check for a finitized moment sequence of each subsystem. If so, go to Step 6; else, set $\xi\rightarrow\xi+1$ and go to Step 2.
6. Stop.

3.3.1 Implementation Issues

Remark 3.3 (Update of denominator estimate)

$\hat{A}_{\xi+1,j}(\bar{p})$: In the iterative process, order of $\hat{A}_{\xi+1,j}(\bar{p})$ increases linearly with ξ. To stop such an increase, the well-known moment-matching technique of model reduction (Jamshidi 1983) is used, where the first $2n_j$ moments of $\frac{1}{\hat{A}_{\xi,j}(\bar{p})\Delta\hat{A}_{\xi,j}(\bar{p})}$ are matched to calculate an n_j-th order $\frac{1}{\hat{A}_{\xi+1,j}(\bar{p})}$.

If the order of the numerator of the jth subsystem is less than n_j-1, it is likely that, as the iterative algorithm converges leading to a finitization of the TMS, Eq. (3.4) may become rank deficient. To circumvent this, for such cases, Eq. (9.4) is modified to (for the ith MISO subsystem),

$$m_{l,j} = \sum_{r=0}^{n_j-1} m_{r,j} a_{l-r,j}; l = 0,1,\cdots$$

Remark 3.4 (Parameter estimation equation)

We have

$$\bar{u}_j(k) = \frac{1}{\hat{A}_{\xi,j}(\bar{p})} u(k); j = 1,2,\ldots,v_i; k \in [N_{\xi-1}, N_\xi], \xi > 1$$

and the parameter estimation equation (in the ξth iteration, $\xi > 1$),

$$\mathcal{F}_{0,d_0+1}(\bar{p})\tilde{y}(k) = \sum_{j=1}^{v_i}\sum_{i=0}^{d_j} m_{i,j}\mathcal{F}_{i,d_0+1}(\bar{p})\frac{1}{\hat{A}_{\xi,j}(\bar{p})} u_j(k).$$

Multiplying by $(\bar{p}+\lambda)^{n_o}$ on both sides, with $n_o = \max_j\{n_j\}$,

$$\beta^{n_o-1}\mathcal{F}_{0,d_0-n_o+1}(\bar{p})\tilde{y}(k) = \sum_{j=1}^{v_i}\sum_{i=0}^{d_j} m_{i,j}\mathcal{F}_{i,d_{0j}+1}(\bar{p})\bar{u}_j'(k),$$

where

$$\bar{u}_j'(k) = \frac{(\bar{p}+\lambda)^{n_j}}{\hat{A}_{\xi,j}(\bar{p})} u_j(k),$$

and

$$d_{0j} = \begin{cases} d_0 & \xi = 1 \\ d_0 - n_o + n_j & \xi > 1 \end{cases}.$$

Thus, we have the reformulated parameter estimation equation (for $\xi > 1$).

$$\tilde{y}^*(k) = \Phi^{\mathrm{T}}(k)\theta,$$

where

$$\tilde{y}^*(k) = \beta^{n_0-1}\mathcal{F}_{0,d_0-n_o+1}(\bar{p})\tilde{y}(k),$$

and

$$\Phi = \left[\mathcal{F}_{0,d_{01}+1}(\bar{p})\bar{u}_1'(k)\cdots\mathcal{F}_{d_1,d_{01}+1}(\bar{p})\bar{u}_1'(k)\middle|\cdots\middle|\mathcal{F}_{0,d_{0v_i}+1}(\bar{p})\bar{u}_{v_i}'(k)\cdots\mathcal{F}_{d_{v_i},d_{0v_i}+1}(\bar{p})\bar{u}_{v_i}'(k)\right]^{\mathrm{T}}.$$

This reformulation has the advantage that the order of the filter $\mathcal{F}$ is reduced considerably for $\xi > 1$, thus requiring less computational effort. Further saving in number of computations is possible by computing $\bar{u}_j'(k)$ as

$$\bar{u}_j'(k) = \frac{1}{\hat{a}_{n_j,j}}\left[u_j(k) - \sum_{i=1}^{n_j} \alpha_i \mathcal{F}_{0,i}(\bar{p})\bar{u}_j'(k)\right], \tag{3.7}$$

where

$$\alpha_i = \frac{1}{\beta_i} \sum_{h=n_j-i}^{n_j} {}^{h}C_{h-n_j+i}(-\lambda)^{h-n_j+1}\hat{a}_{h,j},$$

and

$$\mathcal{F}_{0,i}(\bar{p}) = \sum_{h=0}^{i} {}^{i}C_h(-\lambda)^h \beta^{i-h} \mathcal{F}_{0,d_0-i+h+1}(\bar{p}). \tag{3.8}$$

Thus one may compute outputs of the PFC: $\mathcal{F}_{0,1}(\bar{p}), \mathcal{F}_{0,2}(\bar{p}), \cdots\ldots$ and then compute $\bar{u}_j'(k)$ and the data vector elements using Eqs. (3.7) and (3.8), respectively.

Remark 3.5 (Effect of initial conditions)

For this, the data vector $\Phi(t)$ and the parameter vector θ are augmented by $\left[\mathcal{I}_0(t), \mathcal{I}_1(t),, \mathcal{I}_{d_0-1}(t)\right]^T$ and $\left[c_0(t), c_1(t), \cdots, c_{d_0-1}(t)\right]^T$, respectively, where

$$\mathcal{I}_i(t) = \sum_{h=0}^{i} \frac{t^{d_0-i+h}}{(d_0-i+h)!} {}^{i}C_h(-\lambda)^h,$$

and $c_0, c_1, \cdots, c_{d_0-1}$ are the parameters corresponding to initial conditions.

Remark 3.6 (Check for finitized moment sequence)

For the jth subsystem, the finitization is indicated by $\Delta\hat{A}_{\xi,j}(\bar{p}) \to 1$.

Remark 3.7

Conditions for convergence of this algorithm remain the same as those of the MP case. Proof is omitted here.

3.4 Illustrative Examples

Example 3.1: A SISO system:

Consider a SISO system:

$$G^0(s) = \frac{1.0}{1 + 0.6s + 0.25s^2}.$$

This system is simulated with a PRBS input $T_s = 0.01$. The following cases of noise $v(k)$ contained in measured output $y(k)$ are considered:

1. White noise, $v(k) = w(k)$
2. Colored noise,

$$v(k) = \frac{0.04762 - 0.04762q^{-1}}{1.0 - 0.904762q^{-1}} w(k),$$

where $w(k)$ is zero mean white stationary sequence. The variance of $w(k)$ is adjusted to obtain the desired signal-to-ratio, in both the cases. In parameter estimation, $d = 4, \lambda = 1, \beta = 1$ are taken. Data is obtained for 150 s.

Tables 3.1 and 3.2 show the estimates for white noise and colored noise, respectively, at 10% output noise. Figures 3.1 and 3.2 show the patterns of parameter convergence. In these figures, each 30 s on the time axis corresponds to a single iteration.

TABLE 3.1

Parameter Estimates—Noise: 10% White (Example 3.1)

Iteration Number	$b_0(1.0)$	$a_1(0.6)$	$a_2(0.25)$
1	0.9793	0.6763	0.2591
2	1.0003	0.6138	0.2559
3	0.9994	0.6016	0.2533
4	0.9864	0.6117	0.2674
5	1.0068	0.5972	0.2465

TABLE 3.2

Parameter Estimates—Noise: 10% Colored (Example 3.1)

Iteration Number	$b_0(1.0)$	$a_1(0.6)$	$a_2(0.25)$
1	0.9797	0.6742	0.2556
2	1.0009	0.6152	0.2540
3	0.9997	0.6112	0.2533
4	0.9864	0.6122	0.2671
5	1.0070	0.5974	0.2461

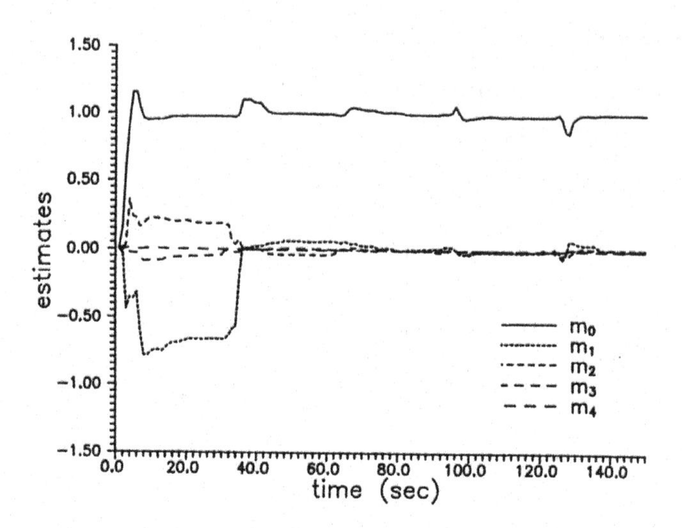

FIGURE 3.1
Pattern of parameter convergence—Noise: 10% white (Example 3.1).

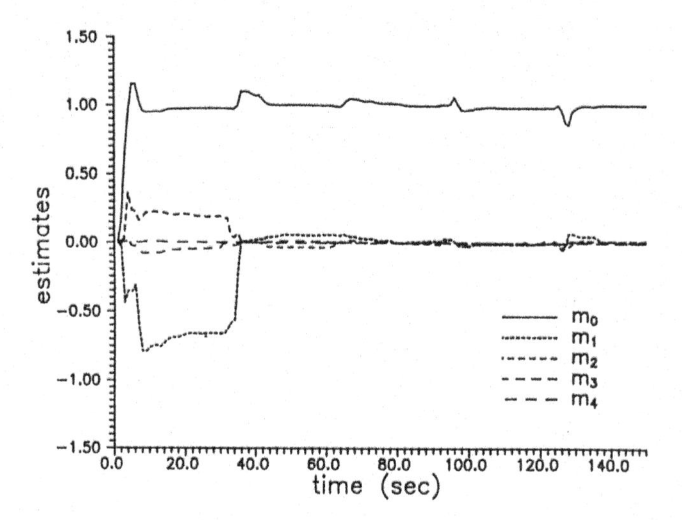

FIGURE 3.2
Pattern of parameter convergence—Noise: 10% colored (Example 3.1).

Example 3.2 (A MISO example):

Reconsider Example 2.3:

$$G^0(s) = \left[\frac{1}{s^2+3s+2} \quad \frac{1}{s+2}\right].$$

In this case, PRBS signals of low mutual correlation are chosen as inputs. Data are obtained for 200 s at $T_s = 0.001$ s. For parameter estimation, $d_1 = 5, d_2 = 3, \lambda = 1, \beta = 1$ are taken.

TABLE 3.3

Parameter Estimates—Noise: 10% White (Example 3.2)

Iteration Number	$b_{0,1}(0.5)$	$a_{1,1}(1.5)$	$a_{2,1}(0.5)$	$b_{0,2}(0.5)$	$a_{1,2}(0.5)$
1	0.5130	0.9782	0.0467	0.5702	0.5807
2	0.4837	1.4693	0.5979	0.5045	0.5215
3	0.5048	1.4588	0.5024	0.5174	0.5219
4	0.4950	1.4748	0.5186	0.5036	0.5048
5	0.5067	1.4806	0.4934	0.5018	0.4846

TABLE 3.4

Parameter Estimates—Noise: 10% Colored (Example 3.2)

Iteration Number	$b_{0,1}(0.5)$	$a_{1,1}(1.5)$	$a_{2,1}(0.5)$	$b_{0,2}(0.5)$	$a_{1,2}(0.5)$
1	0.5133	0.9752	0.0407	0.5701	0.5810
2	0.4838	1.4701	0.5973	0.5045	0.5217
3	0.5050	1.4586	0.5024	0.5177	0.5230
4	0.4952	1.4751	0.5174	0.5037	0.5044
5	0.5068	1.4806	0.4924	0.5018	0.4847

TABLE 3.5

Results of Monte Carlo Simulation Experiments—Noise: White (Example 3.2)

Noise Level (%)	$b_{0,1}(0.5)$	$a_{1,1}(1.5)$	$a_{2,1}(0.5)$	$b_{0,2}(0.5)$	$a_{1,2}(0.5)$	*REN* (%)
10	0.5068 ± 0.0005	1.4863 ± 0.0049	0.4967 ± 0.0039	0.5018 ± 0.0002	0.4875 ± 0.0027	0.88
20	0.5109 ± 0.0007	1.4890 ± 0.0074	0.4844 ± 0.0059	0.5029 ± 0.0003	0.4818 ± 0.0041	1.25
50	0.5201 ± 0.0012	1.4953 ± 0.0114	0.4567 ± 0.0102	0.5055 ± 0.0007	0.4693 ± 0.0066	2.49
100	0.5320 ± 0.0019	1.5032 ± 0.0147	0.4221 ± 0.0146	0.5088 ± 0.0009	0.4537 ± 0.0089	4.21

Tables 3.3 and 3.4 show the iterative estimates with white noise and colored noise, respectively, at 10% output noise obtained at the end of five iterations. The results of Monte Carlo simulation of 20 experiments performed at noise levels ranging from 10% to 100% for the two types of noise sequences considered in Example 3.1 are shown in Tables 3.5 and 3.6, respectively.

Discussion

Iterative finitization of the moment sequence is observed in all cases. As expected, the effect of this finitization is seen in the TFM estimates. See Tables 3.1 through 3.4.

TABLE 3.6

Results of Monte Carlo Simulation Experiments—Noise: Colored (Example 3.2)

Noise Level (%)	$b_{0,1}(0.5)$	$a_{1,1}(1.5)$	$a_{2,1}(0.5)$	$b_{0,2}(0.5)$	$a_{1,2}(0.5)$	*REN* (%)
10	0.5139 ± 0.0005	1.4515 ± 0.0036	0.4250 ± 0.0182	0.5021 ± 0.0007	0.4753 ± 0.0025	4.09
20	0.5228 ± 0.0007	1.4286 ± 0.0059	0.3705 ± 0.0293	0.5033 ± 0.0012	0.4582 ± 0.0039	6.78
50	0.5427 ± 0.0015	1.3789 ± 0.0109	0.2608 ± 0.0524	0.5059 ± 0.0022	0.4204 ± 0.0069	12.35
100	0.5681 ± 0.0023	1.3170 ± 0.0173	0.1451 ± 0.0798	0.5093 ± 0.0035	0.3730 ± 0.0105	18.52

Results of Monte Carlo simulation experiments reveal that the REN is remarkably small at all noise levels (ranging from a very low of 10% to a very high of 100%), and the standard deviation of parameter estimates is insignificant. Results at very high noise levels can be improved further by taking large samples of measurements.

3.5 Choice of Basis of Parameterization

In many practical situations where physical phenomena described by complicated relationships are governed by simplified models, the assumption that a given model set includes a precisely correct model of the observed process is not very realistic. In such cases, modeling error is inevitable and the performance (namely, predictive ability of the estimated models) depends on the choice of model structure and the prior knowledge embedded into the chosen model structures for a given model order.

In the proposed class of linear parameterizations, the following sets of functions are proposed in the previous and the current chapters as the basis for CT system modeling:

1. Motivated by Markov-Poisson parameter models, with a PFC:

$$\mathcal{B}_{\text{PF}} = \left[\frac{\beta}{p+\lambda}, \left(\frac{\beta}{p+\lambda}\right)^2, \ldots, \left(\frac{\beta}{p+\lambda}\right)^d\right]$$

2. Motivated by TM models, with a state-variable filter (SVF):

$$\mathcal{B}_{\text{SVF}} = \left[\frac{1}{E(p)}, \frac{p}{E(p)}, \ldots, \frac{p^{d-1}}{E(p)}\right],$$

where $\frac{1}{E(p)}$ is a dth-order stable filter. A typical choice is a dth-order Poisson filter element,

$$E(p)^{-1} = \left(\frac{\beta}{p+\lambda} \right)^d. \tag{3.9}$$

Unless mentioned otherwise, the same is considered in the rest of this book.

The following issues are now studied via numerical examples:

1. Predictive ability: As pointed out in Remark 3.4 [Eq. (3.8)], $\mathcal{B}_{\text{PF}}$ and $\mathcal{B}_{\text{SVF}}$ are related through a linear nonsingular transformation (for $\lambda \neq 0$), for example, for $d = 4$ and $\beta = 1$,

$$\mathcal{B}_{\text{SVF}} = \begin{bmatrix} 0 & 0 & 0 & 1 \\ 0 & 0 & 1 & -\lambda \\ 0 & 2 & -2\lambda & \lambda^2 \\ 1 & -3\lambda & 3\lambda^2 & \lambda^3 \end{bmatrix} \mathcal{B}_{\text{PF}}.$$

   Hence for a given model order, models based on these two sets will have the same predictive ability.

2. Numerical behavior of the estimation algorithm: This is dictated by the condition number of the matrix

$$\mathbf{R} = \sum_{k=}^{N} \Phi(k)\Phi(k)^{\mathrm{T}}.$$

   It has been pointed out by Subrahmanyam and Rao (1993) that the use of $\mathcal{B}_{\text{PF}}$ results in high condition numbers of the above matrix because these functions are overlapping and non-orthogonal. On the other hand, the second set $\mathcal{B}_{\text{SVF}}$ is near-orthogonal (Goodwin et al. 1991b), which improves the condition number.

3. This situation regarding numerical conditioning may be improved if an intelligently chosen linear transformation of these sets of basis functions is made before parameter estimation commences. When such transformation results in an orthogonal set, the numerical properties of the algorithm will be significantly improved. A popular orthogonal basis is in terms of Laguerre filters:

$$\mathcal{B}_{\text{LAG}} = \left[\frac{1}{p+\lambda}, \frac{1}{p+\lambda}\left(\frac{p-\lambda}{p+\lambda}\right), \ldots, \frac{1}{p+\lambda}\left(\frac{p-\lambda}{p+\lambda}\right)^{d-1} \right]^{\mathrm{T}}.$$

   The required linear transformations are (say, *for* $d = 4$ and $\beta = 1$)

$$\mathcal{B}_{\text{LAG}} = \begin{bmatrix} 1 & 0 & 0 & 0 \\ 1 & -2\lambda & 0 & 0 \\ 1 & -4\lambda & 4\lambda^2 & 0 \\ 1 & -6\lambda & -12\lambda^2 & -8\lambda^3 \end{bmatrix} \mathcal{B}_{\text{PF}},$$

and

$$\mathcal{B}_{\text{LAG}} = \begin{bmatrix} 1 & 3\lambda^2 & 3\lambda & 1 \\ -\lambda^3 & -\lambda^2 & \lambda & 1 \\ \lambda^3 & -\lambda^2 & -\lambda & 0 \\ 1 & 3\lambda^2 & -3\lambda & 1 \end{bmatrix} \mathcal{B}_{\text{SVF}}.$$

Example 3.3

Consider a CT system

$$G^0(s) = \frac{2s+6}{s^3+7s^2+11s+6}.$$

This is simulated using BPFs (an orthogonal set of functions that are used to expand functions of time in orthogonal series. This should not be confused with $\mathcal{B}_{\text{PF}}$ which are transfer functions of Poisson filter segments) approximation with a Gaussian white input of variance of 1.0 and d.c. gain of 1.0 for 100 s at a sampling time of 0.1 s. Models of various orders are estimated with $\beta = \lambda = 1.0$. Parameter estimates are shown in Table 3.7.

TABLE 3.7

Parameter Estimates (Example 3.3)

Basis	d	θ
$\mathcal{B}_{\text{PF}}$	1	[0.9827]
	2	[0.1914, 0.8133]
	3	[0.0585, 1.4174, −0.4760]
	4	[0.0556,1.4320, −0.5038, 0.0163]
	5	[0.0459, 1.5054, −0.7109, 0.2783, −0.1190]
$\mathcal{B}_{\text{SVF}}$	1	[0.9824]
	2	[1.0046, 0.1867]
	3	[0.9999, 1.5233, 0.0562]
	4	[1.0000, 2.5116, 1.5886, 0.0517]
	5	[0.9998, 3.5395, 4.0578, 1.6732, 0.0425]
$\mathcal{B}_{\text{LAG}}$	1	[0.9827]
	2	[0.5981, −0.4066]
	3	[0.6482, −0.4707, −0.1190]
	4	[0.6477, −0.4702, −0.1198, −0.0020]
	5	[0.6482, −0.4718, −0.1179, −0.0050, −0.0074)

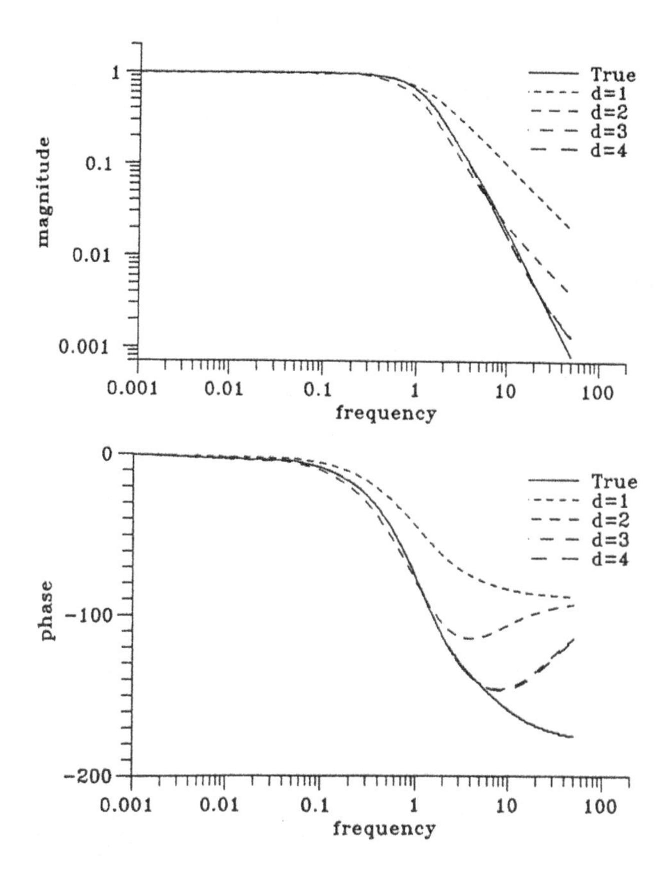

FIGURE 3.3
Results of restricted complexity model estimation using basis $\mathcal{B}_{PF}$ (Example 3.3).

Frequency response plots of the true system and the estimated models are shown in Figures 3.3 through 3.5, respectively, for $\mathcal{B}_{PF}, \mathcal{B}_{SVF}$, and $\mathcal{B}_{LAG}$. As seen from these figures, the estimation quality improves with increase in model order d. Figure 3.6 shows that, for a given model order (say $d = 4$), all the three sets of basis functions produce models with almost the same frequency response. Table 3.8 compares the condition numbers for increasing d. Though the condition numbers with $\mathcal{B}_{PF}$ and $\mathcal{B}_{SVF}$ are larger than those with $\mathcal{B}_{LAG}$, the estimates do not show significant numerical errors (Figure 3.6). However, the same may not be the case with small-word-length computers, and transformations such as suggested in Item 3 above will be helpful.

Example 3.4 (A resonant system):

We now present another example to illustrate an alternative way of modeling resonant systems.

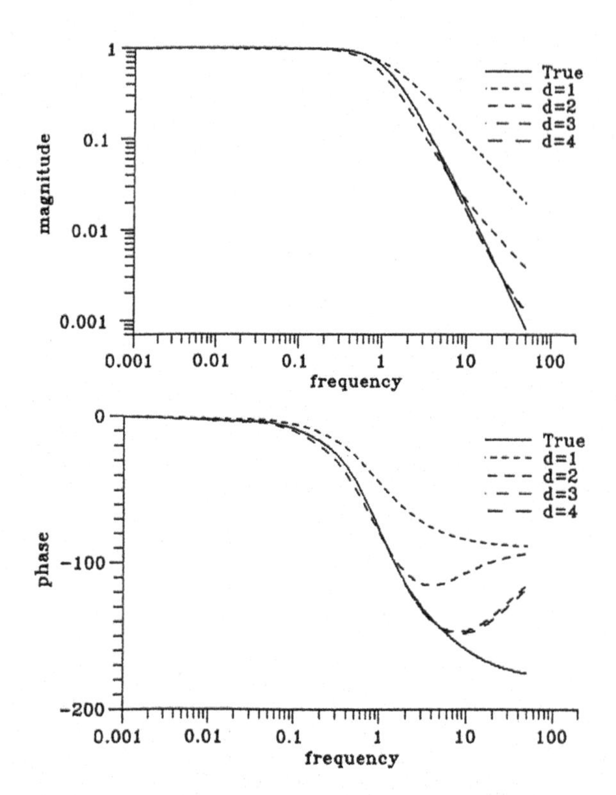

FIGURE 3.4
Results of restricted complexity model estimation using basis $\mathcal{B}_{\text{SVF}}$ (Example 3.3).

Reconsider Example 2.4. A model of the following structure

$$G(p,\theta) = \frac{\sum_{i=1}^{d} m_i p^i}{\left(p^2 + 2\zeta\omega_n p + \omega_n^2\right)^{d/2}}, \quad d \text{ even}$$

with $\zeta = 0.2$ and $\omega_n = 1.2$ (for the original system these are respectively 0.1 and 1.0) is estimated. This may be written in terms of $\mathcal{B}_{\text{SVF}}(p)$ with $E(p)$ of Eq. (3.9) as

$$E(p) = \left(p^2 + 2\zeta\omega_n p + \omega_n^2\right)^{d/2}.$$

This parameterization is in spirit close to the case of Kautz filters. By regrouping terms of these filters, the above parameterization may be obtained.

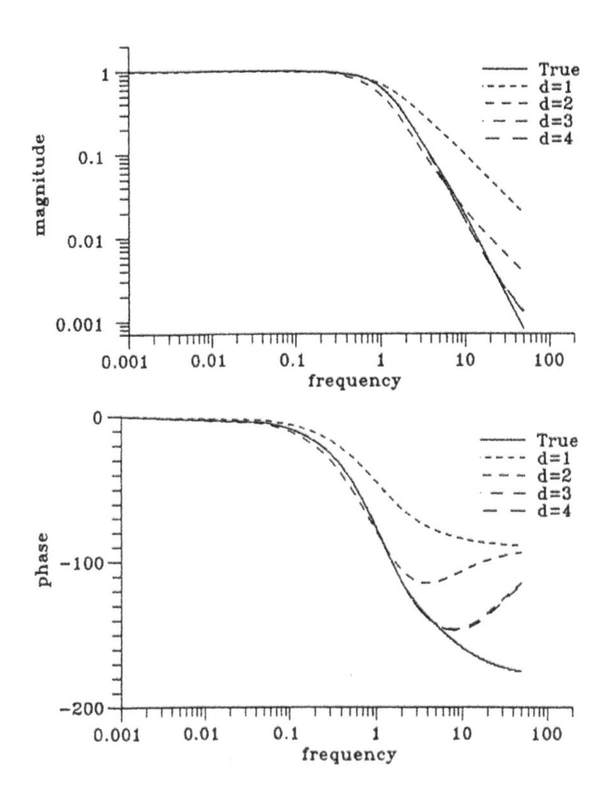

FIGURE 3.5
Results of restricted complexity model estimation using basis $\mathcal{B}_{LAG}$ (Example 3.3).

Models of orders 2, 4, and 6 are estimated and the parameter estimates are shown in Table 3.9. Frequency responses of these estimated models are shown in Figure 3.7. This example shows how prior knowledge of system dynamics may be embedded into the basis to model even complex systems with a small number of parameters.

These illustrative examples motivate the use of $\mathcal{B}_{SVF}$ with the filter polynomial $E(p)$ chosen to represent prior knowledge of poles, whether real or complex, in general as

$$E(p) = \prod_{i=1}^{n_r} (p + \lambda_i)^{n_i} \prod_{i=1}^{n_c} \left(p^2 + as + b\right)^{n_i}.$$

The advantage of regrouping terms of the resultant parameterization in terms of Laguerre/Kautz filters is to have a well-conditioned estimation. Otherwise, they provide the same level of approximation.

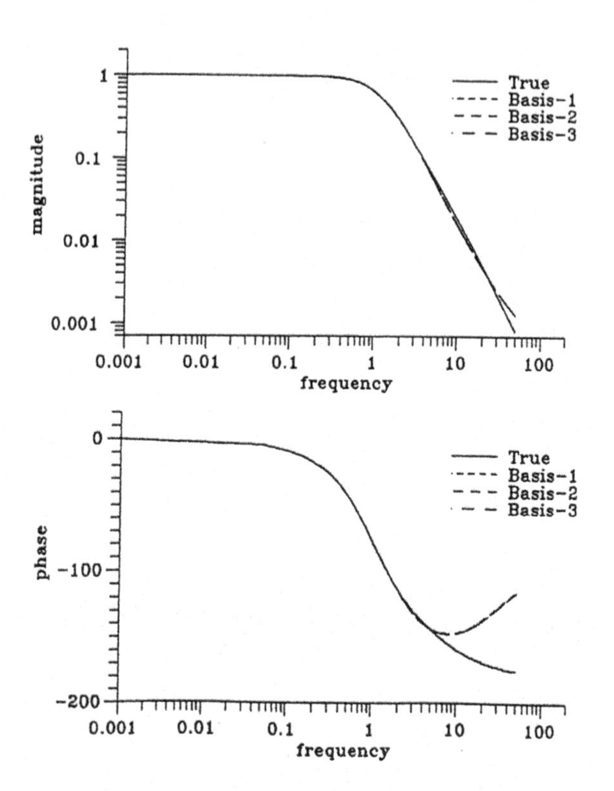

FIGURE 3.6
Comparison of approximations (Example 3.3). Basis-1: $\mathcal{B}_{MP}$, Basis-2: $\mathcal{B}_{SVF}$, Basis-3: $\mathcal{B}_{LAG}$.

TABLE 3.8
Comparison of Condition Numbers (Example 3.3)

d	$\mathcal{B}_{PF}$	$\mathcal{B}_{SVF}$	$\mathcal{B}_{LAG}$
1	1.000000	1.000000	1.000000
2	186.7289	45.64127	45.64652
3	1880.815	159.8336	70.09879
4	11109.74	380.5729	94.07874
5	11698.33	1273.411	120.6597

TABLE 3.9
Parameter Estimates (Example 3.4)

d	θ^T
2	[1.4143, −0.6455]
4	[2.1353, −0.1078, 1.0653, −0.1991]
6	[3.0206, 1.8372, 3.0559, 1.3535, 0.6472, 0.1282]

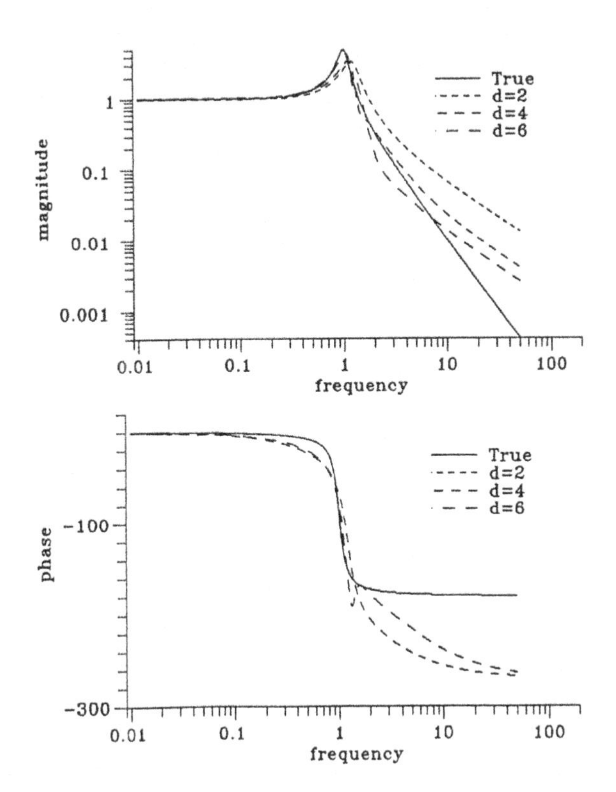

FIGURE 3.7
Results of restricted complexity model estimation using basis $\mathcal{B}_{SVF}$ (Example 3.4).

3.6 Summary and Conclusions

In this chapter, CT models based on time moments of impulse response are studied. Given that the basis by itself is non-realizable, the introduction of a suitable prefilter leads to a realizable identification model. The bias reparameterization algorithm presented here for finitization of TMS is effective in estimating irreducible models of MIMO systems. This is achieved by estimating denominator polynomials of all subsystems and then cancelling them via a suitable filtering of the input signal, in an iterative fashion.

The proposed approach is also suitable for the identification of structural invariants. With the present state–space description, this is equivalent to finding the order of each subsystem. In view of the decoupled estimation model, the order of any subsystem may be found by testing the rank of the associated Hankel matrix formed from the estimated time moments of the subsystem. For this purpose, it is advisable to estimate a sufficiently long sequence of time moments for the sake of accurate order determination.

The bias reparameterization algorithms presented in this and the previous chapters have the same convergence conditions. These algorithms converge only when the Hankel matrix formed using infinite length MPS/TMS has a finite rank, that is, the system should have a finite dimensional structure. This, however, may not always occur in reality. In such cases, one has to be content with whatever estimates are obtained at the end of the first iteration.

By embedding prior knowledge of the modes (real/complex) of the system, it is shown in this chapter that it is possible to estimate good lower order approximations of even complex systems. In such cases, the parameterization with $\mathcal{B}_{SVF}$ as basis seems to be a natural choice.

4

Robust Parameter Estimation

4.1 Introduction

The celebrated LS algorithm, widely used in various problems of parameter estimation, has some useful properties when the probability distributions of the observations are Gaussian. In addition, if the variance of the disturbance sequence, in the absence of undermodeling, is known, the LS solution is also maximum-likelihood one, in which case the estimates are consistent and asymptotically efficient, and the variance of the estimates is computable from the associated covariance matrix of the LS algorithm.

The Gaussian assumption, though not a prerequisite for the derivation of the LS algorithm, helps analyze and explain the statistical properties of the estimate. However, when the Gaussian assumptions are violated, or the modeling errors include systematic components (say, unmodeled dynamics), or statistics of such terms are not available *a priori*, it is difficult to ascertain the performance of the LS estimator because the covariance matrix does not reflect the actual estimation error. (In the $\mathcal{H}_\infty$ literature, the term "disturbance" is used in place of "modeling error." The term modeling error has a broader meaning that typically includes the effects of disturbances, unmodeled dynamics, etc.) This prompts robust estimation algorithms that guarantee small (in energy) estimation errors when the modeling error energy is small, that is, estimators with a bounded $\mathcal{H}_\infty$ norm.

As a dual to the robust control problem proposed by Zames (1981), a complete filtering theory under an $\mathcal{H}_\infty$ performance criterion has been developed (Nagpal and Khargonekar 1991; Shaked and Theodor 1992; Sun et al. 1993). Unlike the Kalman filter theory, filters derived under the $\mathcal{H}_\infty$ theory guarantee an upper bound on the possible estimation error energy. Connections between the $\mathcal{H}_\infty$ filtering algorithms and the Kalman filter have been established by Hassibi et al. (1993b, 1993c).

Inspired by the above work, in this chapter, we consider robustification of the LS algorithm under an $\mathcal{H}_\infty$ performance criterion. As a special case of the $\mathcal{H}_\infty$ filtering, an $\mathcal{H}_\infty$-norm bounded least squares (NBLS) algorithm is derived here for parameter estimation of linear regression models. In $\mathcal{H}_\infty$ filtering, the $\mathcal{H}_\infty$ norm of the filter is assumed to be known *a priori*. Instead, a computational procedure is suggested in this chapter to compute a bound over such a norm from the actual measurement data such that convergence of the estimation algorithm is guaranteed. Implications of this robustification of the LS algorithm in this manner are explored from the point of view of parameter estimation.

To ascertain the quality of the estimate, it is usual to make stochastic assumptions on additive errors before finally computing the variance of the estimate. However, such results may not be valid when computed over a finite amount of data. In this chapter, non-asymptotic formulae are derived to compute ellipsoidal outer bounds that quantify the quality of the estimate if bounds on additive modeling errors are available *a priori*. This makes the NBLS algorithm suitable for deterministic robust estimation.

The layout of this chapter is as follows: In Section 4.2, the problem of parameter estimation under an $\mathcal{H}_\infty$-norm is posed as a special case of filtering. However, because there is no closed-form solution, a suboptimal problem that is equivalent to minimization of a second order criterion function is considered, as suggested by Hassibi et al. (1993b).

The NBLS algorithm is derived in Section 4.3. This algorithm is generalized to include data weighting and a forgetting factor to tackle time-varying systems. A necessary condition for convergence of this algorithm is the convergence of the LS algorithm. It is shown that the NBLS estimate converges to the LS estimate in the asymptotic case provided both the estimates exist. Accordingly, most of the variants of the LS algorithm (e.g., GLS, IV) may be extended with ease.

In Section 4.4, formulae to compute bounds on parameter errors are derived. A comparison with the existing ellipsoidal outer bounding (EOB) algorithm (Fogel and Huang 1982) for SMI, which is an extension of the LS algorithm, is also made. Although the bounds appear to be conservative when compared with the EOB algorithm, it is shown in this section that this conservatism vanishes in the asymptotic case.

Finally, concluding remarks are made in Section 4.5.

4.2 Problem Description

Consider the standard linear regression model of the form:

$$y(k) = \Phi(k)^{\mathrm{T}}\theta + e(k), \tag{4.1}$$

where $\{e(k)\}$ denotes additive modeling errors and $\theta \in \mathbb{R}^d$. The regressor $\Phi(k)$ is formed from the input and output measurement record $\mathbf{Y}^N$. In the case of CT modeling, $\Phi(k)$ is formed from the filtered versions of the elements of $\mathbf{Y}^N$. Assume that the absolute values of the elements of $\mathbf{Y}^N$ and those of $\Phi(k), \forall k$ are bounded.

Given $\mathbf{Y}^N$, the problem is to find an estimator $\Gamma(\mathbf{Y}^N)$ minimizing the infinity norm of a transfer operator T that maps the unknown modeling errors and prior knowledge of θ to estimation errors and then to obtain the resulting infinity norm,

$$\gamma_0 = \operatorname*{Inf}_{\Gamma} \|T\|_\infty$$

and an estimate $\hat{\theta}(\mathbf{Y}^N) = \Gamma(\mathbf{Y}^N)$. For a stable operator T, the $\mathcal{H}_\infty$ norm is defined as

$$\|T\|_\infty = \sup_{\|x\| \neq 0} \frac{\|Tx\|_2}{\|x\|_2}, \tag{4.2}$$

where $\|x\|_2$ is the l_2 norm of a sequence $\{x\}$.

It is well established (Nagpal and Khargonekar 1991; Sun et al. 1993; Hassibi et al. 1993b, 1993c) that estimators based on the above criterion minimize the maximum energy gain from the modeling errors to estimation errors. This will guarantee that if the modeling errors are small (in energy), then the estimation errors will be as small (in energy) as possible, no matter what the modeling errors are or how they are distributed.

To appreciate the implications of this problem, consider the following. Given $\mathbf{Y}^k, k = 1, 2, \cdots, N$, let $\hat{\theta}(k) = \Gamma(\mathbf{Y}^k)$. At the end when $\hat{\theta}(N)$ is available,

$$\left\{ y(k) - \Phi^{\mathrm{T}} \hat{\theta}(N) \right\}$$

is the total modeling error. However, during the estimation, the actual estimation error is

$$\left\{ \Phi^{\mathrm{T}}(k)\hat{\theta}(N) - \Phi^{\mathrm{T}}(k)\hat{\theta}(k) \right\}.$$

The problem considered here is to find such a $\hat{\theta}(N)$ that minimizes the infinity norm of an operator that maps the modeling error energy (and prior knowledge of θ, if available) into estimation error energy.

The problem is thus to find a $\hat{\theta}(N) = \Gamma(\mathbf{Y}^N)$ that minimizes $\|T\|_\infty$ and obtain

$$\gamma_0^2 = \inf_{\Gamma} \|T\|_\infty^2 = \inf_{\Gamma} \sup_{e \in h_2} \frac{\|\tilde{y}\|_2^2}{\left[\theta - \hat{\theta}(0)^{\mathrm{T}}\right] \mathbf{P}(0)^{-1} \left[\theta - \hat{\theta}(0)^{\mathrm{T}}\right] + \|e\|_2^2}, \tag{4.3}$$

where

$$\tilde{y}(k) = \Phi^{\mathrm{T}} \theta(k) - \Phi^{\mathrm{T}} \hat{\theta}(k) \tag{4.4}$$

is the estimation error, and the positive definite matrix $\mathbf{P}(0)$ is a measure of confidence of the prior estimate $\hat{\theta}(0)$ of θ. In the case of the EOB algorithm (Fogel and Huang 1982), this matrix represents the prior parameter set

$$\left\{\theta : \left[\theta - \hat{\theta}(0)\right]^{\mathrm{T}} \mathbf{P}^{-1}(0)\left[\theta - \hat{\theta}(0)\right] \leq 1\right\}.$$

Note that Eq. (4.3) follows from the definition of the $\mathcal{H}_{\infty}$- norm [Eq. (4.2)].

Because of the presence of the unknown γ_0 on the left-hand side of Eq. (4.3), a closed-form solution for θ is not generally possible; hence, the following suboptimal problem is considered (Hassibi et al. 1993b, 1993c).

Problem statement

Given a positive scalar γ, find $\hat{\theta}(N) = \Gamma(\mathbf{Y}^N)$, such that $\|T\|_{\infty} < \gamma$.

From Eq. (4.3), it follows that the above is equivalent to finding a $\hat{\theta}(N)$ such that the following criterion is satisfied:

$$J(\theta) = \left[\theta - \hat{\theta}(0)\right]^{\mathrm{T}} \mathbf{P}^{-1}(0)\left[\theta - \hat{\theta}(0)\right] + \sum_{k=1}^{N}\left[y(k) - \Phi(k)^{\mathrm{T}}\theta\right]^2$$
$$-\gamma^{-2}\sum_{k=1}^{N}\left(\Phi(k)^{\mathrm{T}}\left[\theta - \hat{\theta}(k)\right]\right)^2 > 0 \tag{4.5}$$

This suboptimal $\mathcal{H}_{\infty}$ estimation problem has a solution if and only if $\min_{\theta} J(\theta)$ is positive.

Remark 4.1

As $\gamma \to \infty$, the second-order form Eq. (4.5) reduces to the standard LS criterion function,

$$J(\theta) = \left[\theta - \hat{\theta}(0)\right]^{T} \mathbf{P}^{-1}(0)\left[\theta - \hat{\theta}(0)\right] + \sum_{k=1}^{N}\left[y(k) - \Phi(k)^{T}\theta\right]^2 \tag{4.6}$$

For the LS algorithm, the estimation error energy has, therefore, no finite upper bound.

4.3 Solution to the Suboptimal Problem

We give the solution to the suboptimal $\mathcal{H}_\infty$ estimation problem in the following.

Result 4.1: Norm bounded least squares algorithm

For a given $\gamma > 0$, an estimator Γ with $\|T\|_\infty < \gamma$ || exists if and only if

$$\bar{\bar{\mathbf{P}}}(N)^{-1} = \mathbf{P}(0)^{-1} + \sum_{k=1}^{N} \Phi(k)\Phi(k)^{\mathrm{T}} - \gamma^{-2} \sum_{k=1}^{N-1} \Phi(k)\Phi(k)^{\mathrm{T}} > 0, \tag{4.7}$$

where

$$\bar{\bar{\mathbf{P}}}(N)^{-1} = \bar{\mathbf{P}}(N)^{-1} - \gamma^{-2}\Phi(N)\Phi(N)^{\mathrm{T}}, \tag{4.8}$$

and

$$\bar{\mathbf{P}}(N)^{-1} = \mathbf{P}(0)^{-1} + (1-\gamma^{-2}) \sum_{k=1}^{N} \Phi(k)\Phi(k)^{\mathrm{T}}, \tag{4.9}$$

which satisfies the following recursive equation:

$$\bar{\mathbf{P}}(k) = \bar{\mathbf{P}}(k-1) - (1-\gamma^{-2}) \frac{\bar{\mathbf{P}}(k-1)\Phi(k)\Phi(k)^{\mathrm{T}}\bar{\mathbf{P}}(k-1)}{1-(1-\gamma^{-2})\Phi(k)^{\mathrm{T}}\bar{\mathbf{P}}(k-1)\Phi(k)}. \tag{4.10}$$

Then the parameter estimate $\hat{\theta}(k)$ may be computed recursively as

$$\hat{\theta}(k) = \hat{\theta}(k-1) + \frac{\bar{\mathbf{P}}(k-1)\Phi(k)}{1+\Phi(k)^{\mathrm{T}}\bar{\mathbf{P}}(k-1)\Phi(k)} \bar{\varepsilon}(k), \tag{4.11}$$

where

$$\bar{\varepsilon}(k) = y(k) - \Phi^{\mathrm{T}}(k)\hat{\theta}(k-1). \tag{4.12}$$

Proof

Note that this result is a special case of the well-known $\mathcal{H}_\infty$ filter equations formulated, for example, in Nagpal and Khargonekar (1991) and Hassibi et al. (1993b, 1993c). This result may also be derived as follows.

From Eq. (4.5), an estimator with a bounded $\mathcal{H}_\infty$-norm exits if and only if $\min_\theta J(\theta) > 0$. Accordingly, minimizing $J(\theta)$ with respect to θ, at $\theta = \hat{\theta}(N)$,

$$\bar{\mathbf{P}}(N)^{-1}\hat{\theta}(N) = \bar{\mathbf{P}}(0)^{-1}\hat{\theta}(0) + \sum_{k=1}^{N}\Phi(k)y(k) - \gamma^{-2}\sum_{k=1}^{N}\Phi(k)\Phi(k)^{\mathrm{T}}\hat{\theta}(k). \tag{4.13}$$

Collecting terms of $\hat{\theta}(N)$ on both sides,

$$\bar{\bar{\mathbf{P}}}(N)^{-1}\hat{\theta}(N) = \mathbf{P}(0)^{-1}\hat{\theta}(0) + \sum_{k=1}^{N}\Phi(k)y(k) - \gamma^{-2}\sum_{k=1}^{N-1}\Phi(k)\Phi(k)^{\mathrm{T}}\hat{\theta}(k).$$

For the existence of a unique $\hat{\theta}(N)$, $\bar{\bar{\mathbf{P}}}(N)$ should be positive definite, leading to the convergence condition, Eq. (4.7).

From Eqs. (4.7) and (4.8), Eq. (4.9) follows, which may then be recursified to

$$\bar{\mathbf{P}}(k)^{-1} = \bar{\mathbf{P}}(k-1)^{-1} + (1-\gamma^{-2})\sum_{k=1}^{N}\Phi(k)\Phi(k)^{\mathrm{T}}, \; k = 1,\ldots,N, \tag{4.14}$$

with $\bar{\mathbf{P}}(0) = \mathbf{P}(0)$, or using the matrix-inversion lemma, to Eq. (4.10).

From Eq. (4.13), for some $\mathbf{Y}^{k-1}$ and $\mathbf{Y}^{k}$, we have

$$\bar{\mathbf{P}}(k-1)^{-1}\hat{\theta}(k-1) = \mathbf{P}(0)^{-1}\hat{\theta}(0) + \sum_{i=1}^{k-1}\Phi(i)y(i) - \gamma^{-2}\sum_{i=1}^{k-1}\Phi(i)\Phi(i)^{\mathrm{T}}\hat{\theta}(i), \tag{4.15}$$

$$\bar{\mathbf{P}}(k)^{-1}\hat{\theta}(k) = \mathbf{P}(0)^{-1}\hat{\theta}(0) + \sum_{i=1}^{k-1}\Phi(i)y(i) - \gamma^{-2}\sum_{i=1}^{k-1}\Phi(i)\Phi(i)^{\mathrm{T}}\hat{\theta}(i). \tag{4.16}$$

Subtracting Eq. (4.15) from Eq. (4.16), and using Eqs. (4.14) and (4.12), we get

$$\hat{\theta}(k) = \hat{\theta}(k-1) + \left[\bar{\mathbf{P}}(k-1)^{-1} + \Phi(k)\Phi(k)^{\mathrm{T}}\right]^{-1}\Phi(k)\bar{\varepsilon}(k)$$

to give Eq. (4.11) using matrix inversion lemma.

QED

Result 4.2

Lower bound on γ

For the linear regression model (4.1) and $\mathbf{Y}^N$*, the following conditions are sufficient for the NBLS algorithm to be convergent:*

$$\max_i \lambda_i \left\{ \mathbf{R}(N-1) - \mathbf{P}(0)^{-1} \right\} > 0 \tag{4.17}$$

and

$$\mathbf{R}(N) > 0 \tag{4.18}$$

$$\gamma > \sqrt{\frac{\max_i \lambda_i \left\{ \mathbf{R}(N-1) - \mathbf{P}(0)^{-1} \right\}}{\min_i \lambda_i \left\{ \mathbf{R}(N) \right\}}} \geq 1, \tag{4.19}$$

where

$$\mathbf{R}(N) = \mathbf{P}(0)^{-1} + \sum_{k=1}^{N} \Phi(k)\Phi(k)^{\mathrm{T}}. \tag{4.20}$$

Proof

From Condition (4.7),

$$\left.\begin{aligned} & \mathbf{R}(N) - \gamma^{-2}\left[\mathbf{R}(N-1) - \mathbf{P}(0)^{-1}\right] > 0 \\ & \Rightarrow \gamma^{-2}\left[\mathbf{R}(N-1) - \mathbf{P}(0)^{-1}\right] < \mathbf{R}(N) \\ & \Rightarrow \gamma^{-2} \max_i \lambda_i \left\{ \mathbf{R}(N-1) - \mathbf{P}(0)^{-1} \right\} < \min_i \lambda_i \left\{ \mathbf{R}(N) \right\} \\ & \Rightarrow \gamma > \sqrt{\frac{\max_i \lambda_i \left\{ \mathbf{R}(N-1) - \mathbf{P}(0)^{-1} \right\}}{\min_i \lambda_i \left\{ \mathbf{R}(N) \right\}}} \end{aligned}\right\}. \tag{4.21}$$

The case of $\gamma < 1$ is not acceptable because, when $\gamma < 1$, there may be some N after which Eq. (4.7) is not satisfied. This argument together with Eq. (4.21) gives Eq. (4.19).

Condition (4.18) implies the convergence of the RLS algorithm, that is,

$$\min_i \lambda_i \left\{ \mathbf{R}(N) \right\} > 0.$$

See, for example, Goodwin and Sin (1984) for a proof. Therefore, for γ to be real,

$$\max_i \lambda_i \left\{ \mathbf{R}(N-1) - \mathbf{P}(0)^{-1} \right\} > 0.$$

QED

Result 4.3: Asymptotic convergence to the RLS solution

Given the linear regression model (4.1) and $\mathbf{Y}^N$, *let the RLS algorithm be asymptotically convergent and* $\hat{\theta}_{\mathrm{RLS}}(N)$ *be the estimate for some N. Also let an estimator* Γ *with* $\|T\|_\infty < \gamma$ |, γ *finite exist, giving* $\hat{\theta}(N)$. *Then*

$$\lim_{N\to\infty}\left[\hat{\theta}(N) - \hat{\theta}_{\mathrm{RLS}}(N)\right] = 0. \tag{4.22}$$

Proof

$\hat{\theta}(N)$ and $\hat{\theta}_{\mathrm{RLS}}(N)$ are given by

$$\left.\begin{aligned}&\left[\mathbf{P}(0)^{-1} + \sum_{k=1}^{N}\Phi(k)\Phi(k)^T - \gamma^{-2}\sum_{k=1}^{N-1}\Phi(k)\Phi(k)^T\right]\hat{\theta}(N) = \mathbf{P}(0)^{-1}\hat{\theta}(0),\\&+\sum_{k=1}^{N}\Phi(k)\mathbf{y}(k) - \gamma^{-2}\sum_{k=1}^{N-1}\Phi(k)\Phi(k)^T\hat{\theta}(k)\end{aligned}\right. \tag{4.23}$$

and

$$\left[\mathbf{P}(0)^{-1} + \sum_{k=1}^{N}\Phi(k)\Phi(k)^T\right]\hat{\theta}_{\mathrm{RLS}}(N) = \mathbf{P}(0)^{-1}\hat{\theta}(0) + \sum_{k=1}^{N}\Phi(k)y(k), \tag{4.24}$$

respectively. Note that Eq. (4.24) may be obtained from Eq. (4.23) by setting $\gamma = \infty$. Subtracting Eq. (4.24) from Eq. (4.23), and using Eq. (4.20),

$$\left.\begin{aligned}\mathbf{R}(N)\left[\hat{\theta}(N) - \hat{\theta}_{\mathrm{RLS}}(N)\right] &= \gamma^{-2}\sum_{k=1}^{N-1}\Phi(k)\Phi(k)^{\mathrm{T}}\left[\hat{\theta}(N) - \hat{\theta}(k)\right]\\&= \gamma^{-2}\sum_{k=1}^{N-1}\Phi(k)\tilde{y}(k)\end{aligned}\right\}. \tag{4.25}$$

where $\tilde{y}(k)$ is the estimation error defined in Eq. (4.4) at $\hat{\theta} = \hat{\theta}(N)$.

A necessary condition for the asymptotic convergence of the RLS algorithm is (from Goodwin and Sin 1984)

$$\lim_{N\to\infty}\min_i \lambda_i\{\mathbf{R}(N)\} = \infty. \tag{4.26}$$

The right-hand side of Eq. (4.25) is finite because, when γ is finite, the estimation error energy and consequently the estimation error $\tilde{y}(k)$ is finite, and the elements of $\Phi(k)$ are bounded [see the assumption below Eq. (4.1)]. This argument, together with Eq. (4.26) proves Eq. (4.22).

QED

Remark 4.2

As $\gamma \to \infty$*, the NBLS algorithm reduces to the usual RLS algorithm,*

$$\bar{\mathbf{P}}_{\mathrm{RLS}}(k) = \bar{\mathbf{P}}_{\mathrm{RLS}}(k-1) + \frac{\bar{\mathbf{P}}_{\mathrm{RLS}}(k-1)\Phi(k)\Phi(k)^{\mathrm{T}}\bar{\mathbf{P}}_{\mathrm{RLS}}(k-1)}{1+\Phi(k)^{\mathrm{T}}\bar{\mathbf{P}}_{\mathrm{RLS}}(k-1)\Phi(k)}$$

$$\hat{\theta}(k) = \hat{\theta}(k-1) + \frac{\bar{\mathbf{P}}_{\mathrm{RLS}}(k-1)\Phi(k)}{1+\Phi(k)^{\mathrm{T}}\bar{\mathbf{P}}_{\mathrm{RLS}}(k-1)\Phi(k)}\bar{\varepsilon}(k)$$

where

$$\bar{\varepsilon}_{\mathrm{RLS}}(k) = y(k) - \Phi(k)^{\mathrm{T}}\hat{\theta}_{\mathrm{RLS}}(k-1).$$

This algorithm is known for its good convergence.

Remark 4.3

The case of $\gamma = 1$ *corresponds to the normalized least mean squares (NLMS) algorithm (Widrow and Stearns 1985),*

$$\hat{\theta}_{\mathrm{NLMS}}(k) = \hat{\theta}_{\mathrm{NLMS}}(k-1) - \frac{\bar{\mathbf{P}}(0)\Phi(k)}{1+\Phi(k)^{\mathrm{T}}\bar{\mathbf{P}}(0)\Phi(k)}\bar{\varepsilon}(k)$$

$$\bar{\varepsilon}_{\mathrm{NLMS}}(k) = y(k) - \Phi(k)^{\mathrm{T}}\hat{\theta}_{\mathrm{NLMS}}(k-1)$$

The $\mathcal{H}_\infty$ *optimality of this algorithm has been shown by Hassibi et al. (1993a). For more detailed interpretations see this reference. This algorithm is commonly used for adaptive estimation because of its good tracking ability.*

Remark 4.4

For values of γ *between* 1 *and* 0*, the NBLS algorithm is, therefore, expected to show a performance trade-off between good convergence (of the RLS algorithm) and good tracking (of the NLMS algorithm).*

To further understand the implications of the robust performance criterion, compare

$$\bar{\mathbf{P}}(k) = \bar{\mathbf{P}}(k-1) + \left(1-\gamma^{-2}\right)\Phi(k)\Phi(k)^{\mathrm{T}}$$

and

$$\bar{\mathbf{P}}_{\mathrm{RLS}}(k) = \bar{\mathbf{P}}_{\mathrm{RLS}}(k-1) + \left(1-\gamma^{-2}\right)\Phi(k)\Phi(k)^{\mathrm{T}}.$$

For $\gamma > 1$, given that $(1-\gamma^{-2}) < 1$, the information added to $\bar{\mathbf{P}}(k)$ is less than that of $\bar{\mathbf{P}}_{RLS}(k)$. The NBLS algorithm is therefore conservative or cautious.

We also have

$$\det\{\bar{\mathbf{P}}(N)\} = \frac{\det\{\mathbf{P}(0)\}}{1+(1-\gamma^2)\sum_{k=1}^{N}\Phi(k)^{\mathrm{T}}\mathbf{P}(0)\Phi(k)},$$

and

$$\det\{\bar{\mathbf{P}}_{RLS}(N)\} = \frac{\det\{\mathbf{P}(0)\}}{1+\sum_{k=1}^{N}\Phi(k)^{\mathrm{T}}\mathbf{P}(0)\Phi(k)}.$$

Therefore $\det\{\bar{\mathbf{P}}(N)\} > \det_{RLS}\{\bar{\mathbf{P}}(N)\}$, implying that the decay of the gain of the NBLS algorithm is slower than that of the RLS algorithm.

The following are further implications:

- *In some cases of RLS estimation, the algorithm gain falls rapidly as the covariance matrix becomes weak after a few recursions (Goodwin and Sin 1984), leading to failure of convergence. The NBLS algorithm prevents this in a systematic way because of its cautious $\bar{\mathbf{P}}$ update.*
- *By definition, the NBLS algorithm attempts to converge to a $\hat{\theta}(N)$ such that $\sum_{k=1}^{N}(\Phi^{\mathrm{T}}(k)[\theta-\hat{\theta}(k)])^2$ is as small as possible. This is equivalent to minimizing the absolute value of $[\hat{\theta}(N)-\hat{\theta}(k)]$ weighted by $\Phi(k), \forall k$. During the implementation, $\hat{\theta}(k)$ therefore approaches $\hat{\theta}(N)$ rapidly. This accelerated estimation may be useful when the measurement data records are short. Although the pattern of parameter convergence may be oscillatory, it will eventually converge to the LS estimate as $N \to \infty$. However, the measurement data should allow for a γ small enough for these features to be significant.*

Remark 4.5

In the actual implementation of the NBLS algorithm, it is possible to compute γ recursively while preserving convergence of the algorithm. From Result 4.2,

$$\gamma(k) = \alpha\sqrt{\frac{\max_i \lambda_i\{\mathbf{R}(k-1)-\mathbf{P}(0)^{-1}\}}{\min_i \lambda_i\{\mathbf{R}(k)\}}}, \tag{4.27}$$

where α is a constant chosen to be greater than unity (say, 1.1). However, for small k, conditions (4.17) and (4.18) may not be satisfied. During the implementation, the NBLS algorithm may, therefore, be switched on after processing the data by the RLS algorithm for the first few recursions.

Example 4.1

Consider the following continuous system (Wahlberg and Ljung 1986):

$$G^0(s) = \frac{0.00283s^3 + 0.01893s^2 + 0.0562887s + 0.076}{s^4 + 3.17s^3 + 3.06s^2 + 0.87s + 0.076}.$$

This is simulated with a sampling time $T_s = 0.1$ for 100 s such that $N = 1000$. For parameter estimation, a second-order model under $\mathcal{B}_{SVF}$ with $\lambda = \beta = 5.0$ is chosen. There is no disturbance corrupting the measurements, that is, $e(k)$ represents unmodeled dynamics alone.

Figure 4.1 shows the results of parameter estimation, where the patterns of parameter convergence are plotted. Results of both the LS (the dashed line) and NBLS (the continuous line) algorithms are shown. The NBLS algorithm is switched on after processing the data by the LS algorithm for the first 100 recursions. This is necessary to compute an initial γ, which is updated recursively using Eq. (4.27). The NBLS estimate is obtained at $\gamma = 1.145$ as

$$G\left(s, \hat{\theta}\right) = \frac{-0.0427s + 23.8182}{(s+5)^2},$$

and the RLS estimate is obtained as

$$G\left(s, \hat{\theta}_{RLS}\right) = \frac{-0.04488s + 21.5534}{(s+5)^2}.$$

Notice the slow convergence of the RLS estimation. The estimate of the steady-state gain of the system (1.0) has almost approached (at $k = 500$) a value close to 1.0 (on average) in the case of NBLS estimation, whereas the corresponding RLS estimate tends to approach the same very slowly but smoothly. This is what is meant by accelerated estimation by the NBLS algorithm in Remark 4.4.

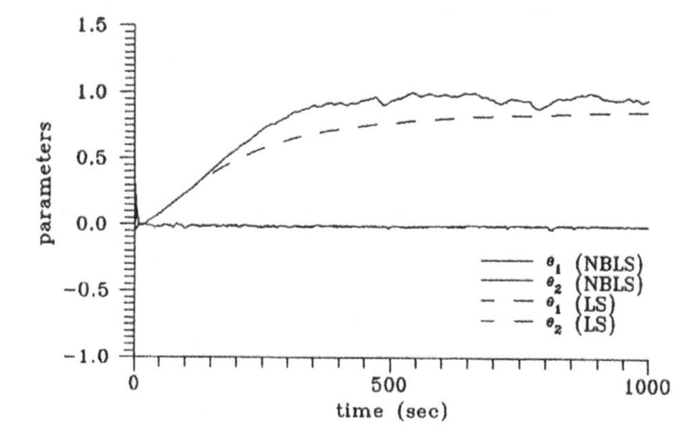

FIGURE 4.1
Patterns of parameter convergence.

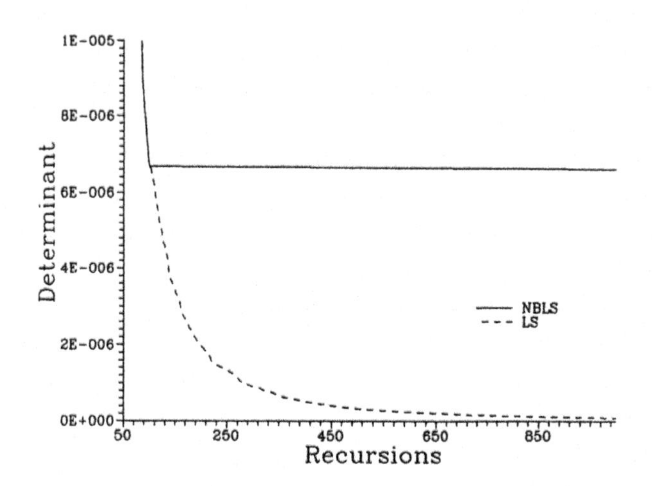

FIGURE 4.2
Decay of determinants.

Figure 4.2 shows the pattern of decay of the determinants of $\bar{\mathbf{P}}$ and $\bar{\mathbf{P}}_{RLS}$. The decay of det($\bar{\mathbf{P}}$) is almost arrested when the NBLS algorithm is switched on at $k = 100$. Although these values are of the order of 10^{-6}, they are strong enough to keep the parameter estimation active, as shown in Figure 4.1.

Figure 4.3 shows the responses of NBLS and RLS algorithms for a sudden change in the steady-state gain of the system from 1.0 to 1.5 at $k = 500$. The NBLS algorithm is able to track this sudden change as indicated by the pattern of parameter θ_1, which corresponds to the steady-state gain. On the other hand, the reaction of the RLS algorithm is very sluggish.

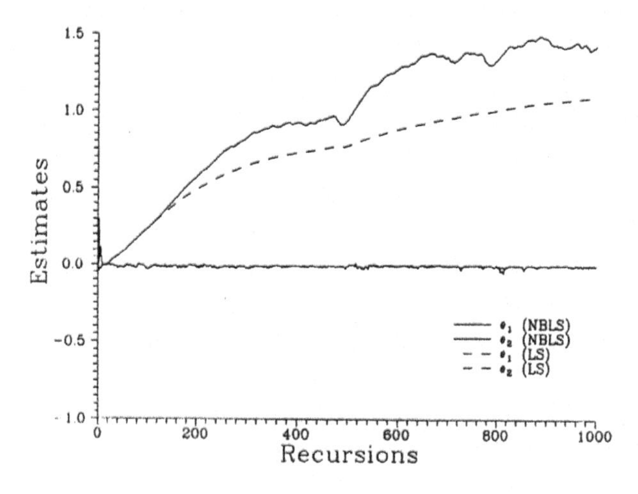

FIGURE 4.3
Response to a sudden change in steady-state gain of the system.

Note that the oscillatory nature of the NBLS parameter pattern is not due to any random error term (the error is deterministic) but is, instead, a property of the estimator. Low-pass filtering of the parameters is required to keep the pattern smooth.

To tackle real-time identification problems, different measurements could be assigned different weights and a forgetting factor in the NBLS criterion function [Eq. (4.5)], that is,

$$J(\theta) = \left[\theta - \hat{\theta}(0)\right]^{\mathrm{T}} \rho^N \mathbf{P}(0)^{-1}\left[\theta - \hat{\theta}(0)\right] + \sum_{k=1}^{N} \alpha_k \rho^{N-k}\left[y(k) - \Phi(k)^{\mathrm{T}}\theta\right]^2 \\ -\gamma^{-2}\sum_{k=1}^{N} \alpha_k \rho^{N-k}\left[\Phi(k)^{\mathrm{T}}\left\{\theta - \hat{\theta}(k)\right\}\right]^2, \tag{4.28}$$

where $\rho < 1$ is known as the forgetting factor, and α_k is a time-varying weight given to the measured data. The solution is summarized below.

Result 4.4: Weighted NBLS algorithm with a forgetting factor

For a given $\gamma > 0$, *an estimator* Γ *with* $J(\theta)$ *defined in* Eq. (4.28) *positive, exists if and only if*

$$\bar{\bar{\mathbf{P}}}(N) = \rho^N \mathbf{P}(0)^{-1} + \sum_{k=1}^{N} \alpha\rho^{N-k}\Phi(k)\Phi(k)^{\mathrm{T}} - \gamma^{-2}\sum_{k=1}^{N-1} \alpha\rho^{N-k}\Phi(k)\Phi(k)^{\mathrm{T}} > 0,$$

where

$$\bar{\bar{\mathbf{P}}}(N) = \bar{\mathbf{P}}(N)^{-1} - \gamma^{-2}\alpha_N\Phi(N)\Phi(N)^{\mathrm{T}}, \tag{4.29}$$

and

$$\bar{\mathbf{P}}(N)^{-1} = \rho^N \mathbf{P}(0)^{-1} + \left(1-\gamma^{-2}\right)\sum_{k=1}^{N} \alpha_k \rho^{N-k}\Phi(k)\Phi(k)^{\mathrm{T}}, \tag{4.30}$$

which satisfies the following recursive equation:

$$\bar{\mathbf{P}}(k) = \frac{1}{\rho}\left\{\rho^N \mathbf{P}(k-1) - \alpha_k\left(1-\gamma^{-2}\right)\frac{\bar{\mathbf{P}}(k-1)\Phi(k)\Phi(k)^{\mathrm{T}}\bar{\mathbf{P}}(k-1)}{\rho + \alpha_k\left(1-\gamma^{-2}\right)\Phi(k)^{\mathrm{T}}\bar{\mathbf{P}}(k-1)\Phi(k)}\right\}.$$

Then $\hat{\theta}(k)$ *may be computed recursively as*

$$\hat{\theta}(k) = \hat{\theta}(k-1) + \frac{\alpha_k \bar{\mathbf{P}}(k-1)\Phi(k)}{\rho + \alpha_k \Phi(k)^{\mathrm{T}} \bar{\mathbf{P}}(k-1)\Phi(k)} \bar{\varepsilon}(k),$$

Proof, being similar to that of Result 4.1, is omitted here. Note that the update equations as given above collapse to those of the ordinary NBLS algorithm when $\rho = 1$ and $\alpha_k = 1, \forall k$.

4.4 Bounds on Parameter Errors

In this section, we give formulae for computing deterministic ellipsoidal bounds on parameter errors. Considering the linear regression model [Eq. (4.1)], assume that the following time-varying bounds are available on the modeling error term $e(k), \forall k$,

$$e(k)^2 \leq r_k. \tag{4.31}$$

Also assume that the prior parameter error bound (or, prior parameter set in SMI parlance) is given as

$$\left[\theta - \hat{\theta}(0)\right] \mathbf{P}(0)^{-1} \left[\theta - \hat{\theta}(0)\right]^{\mathrm{T}} \leq 1. \tag{4.32}$$

We now compute posterior bounds of the form

$$\left[\theta - \hat{\theta}(N)\right] \mathbf{P}(N)^{-1} \left[\theta - \hat{\theta}(N)\right]^{\mathrm{T}} \leq 1.$$

Result 4.5: Ellipsoidal error bounds

Given priors (4.31) *and* (4.32) *along with the measured data* $\mathbf{Y}^N$, *let there exist an estimator* Γ *with* $\|T\|_\infty < \gamma$, *yielding an estimate* $\hat{\theta}(N)$. *Then the posterior error bounds are given by*

$$\left[\theta - \hat{\theta}(N)\right]^{\mathrm{T}} \mathbf{P}(N)^{-1} \left[\theta - \hat{\theta}(N)\right] \leq 1, \tag{4.33}$$

where

$$\mathbf{P}(N)=\left[\bar{\mathbf{P}}(N)^{-1}-\gamma^{-2}\sum_{k=1}^{N-1}\Phi(k)\Phi(k)^{\mathrm{T}}\right]^{-1}\sigma(N), \tag{4.34}$$

$$\left.\begin{aligned}&\sigma(N)=1+\sum_{k=1}^{N}r_k-\sum_{k=1}^{N}\varepsilon(k)^2-\left[\hat{\theta}(N)-\hat{\theta}(0)\right]^{\mathrm{T}}\mathbf{P}(0)^{-1}\left[\hat{\theta}(N)-\hat{\theta}(0)\right],\\&\varepsilon(k)=y(k)-\Phi(k)^{\mathrm{T}}\hat{\theta}(N).\end{aligned}\right\} \tag{4.35}$$

Proof

Let,

$$\bar{\sigma}(N)=\left[\theta-\hat{\theta}(N)\right]^{\mathrm{T}}\left[\bar{\mathbf{P}}(N)^{-1}-\gamma^{-2}\sum_{k=1}^{N-1}\Phi(k)\Phi(k)^{\mathrm{T}}\right]\left[\theta-\hat{\theta}(N)\right]. \tag{4.36}$$

After some manipulations,

$$\begin{aligned}\bar{\sigma}(N)=&\left[\theta-\hat{\theta}(0)\right]^{\mathrm{T}}\mathbf{P}(0)^{-1}\left[\theta-\hat{\theta}(0)\right]+\sum_{k=1}^{N}\left[e(k)^2-\varepsilon(k)^2\right]\\&-\left[\hat{\theta}(N)-\hat{\theta}(0)\right]^{\mathrm{T}}\mathbf{P}(0)^{-1}\left[\hat{\theta}(N)-\hat{\theta}(0)\right].\end{aligned}$$

From priors (4.31), (4.32), and (4.35), we have the following upper bound on $\bar{\sigma}(N)$

$$\bar{\sigma}(N)\leq\sigma(N). \tag{4.37}$$

Substituting Eqs. (4.35) and (4.36) in Eq. (4.37), and using Eq. (4.34), we get Eq. (4.33).

QED

Remark 4.6

Compare the above with the EOB algorithm of Fogel and Huang (1982). By considering a weighted LS criterion function, the EOB algorithm computes the minimal posterior parameter error bounds by choosing the weights $\{\alpha_k\}$ to minimize determinant or trace of the matrix $\mathbf{P}(N)$. Fogel and Huang (1982) (see also Norton 1987b; Wahlberg and Ljung 1992) give recursive formulae to compute α_k analytically.

However, in the case of a weighted NBLS algorithm, the analytical expressions seem to be very complex. As suggested by Wahlberg and Ljung (1982), one possibility is to perform a global numerical optimization to find an optimal sequence $\{\alpha_k\}$ *minimizing the determinant or trace of* $\mathbf{P}(N)$.

Remark 4.7

We have

$$\bar{\mathbf{P}}(N)^{-1}+\gamma^{-2}\sum_{k=1}^{N-1}\Phi(k)\Phi(k)^{\mathrm{T}}=\mathbf{P}(0)^{-1}+\sum_{k=1}^{N}\Phi(k)\Phi(k)^{\mathrm{T}}.$$

Therefore, from Eq. (4.34) ,

$$\det\{\mathbf{P}(N)\}=\frac{\sigma(N)\det\{\mathbf{P}(0)\}}{1+\sum_{k=1}^{N}\Phi(k)^{\mathrm{T}}\mathbf{P}(0)\Phi(k)},$$

and similarly for the LS criterion,

$$\det\{\mathbf{P}_{\mathrm{RLS}}(N)\}=\frac{\sigma_{\mathrm{RLS}}(N)\det\{\mathbf{P}(0)\}}{1+\sum_{k=1}^{N}\Phi(k)^{\mathrm{T}}\mathbf{P}(0)\Phi(k)},$$

where $\sigma_{\mathrm{RLS}}(N)$ *is* $\sigma(N)$ *at* $\gamma=\infty$, *so that*

$$\frac{\det[\mathbf{P}(0)]}{\det[\mathbf{P}_{\mathrm{RLS}}(N)]}=\frac{\sigma(N)}{\sigma_{\mathrm{RLS}}(N)}.$$

In the limit as $N\to\infty$, *given that* $[\sigma(N)-\sigma_{\mathrm{RLS}}(N)]\to 0$, *and* $\sigma(N)\to\sigma_{\mathrm{RLS}}(N)$, *and therefore the above ratio approaches unity. Thus, the conservatism in* $\mathbf{P}$, *in the asymptotic case, does not lead to conservative error bounds.*

4.5 Summary and Conclusions

An $\mathcal{H}_\infty$-norm bounded LS algorithm for robust parameter estimation of linear regression models in a deterministic framework is proposed in this chapter by extending some results of $\mathcal{H}_\infty$ filtering theory. By definition, the proposed algorithm guarantees estimates with the smallest possible estimation error energy over all possible modeling errors of fixed energy; therefore, the algorithm is robust. As a result of the underlying $\mathcal{H}_\infty$ performance criterion, the NBLS estimator is conservative. This leads to active and accelerated parameter estimation.

Here, the achievable level of $\mathcal{H}_\infty$-norm is dictated by the experimental conditions such as model structure, and input signals [Eq. (4.19)]. With these, an explicit procedure has been proposed in this chapter to compute the norm. It has been shown that the NBLS estimate converges to the RLS estimate in the asymptotic case. Intuitively, it is therefore expected that all asymptotic results (such as conditions of persistency of excitation, frequency domain interpretations, and frequency domain criteria for experiment design) available for the RLS algorithm will remain valid for the NBLS algorithm.

In general, variance estimates may be misleading when computed over a finite amount of data. In such cases, non-asymptotic deterministic formulae such as those given here provide valid bounds. These formulae are also useful for the exercise of error quantification (to be treated in the next chapter). Note that the conservatism of NBLS recursions and the computed bounds vanishes in the asymptotic case (see Remark 4.7).

5

Error Quantification

5.1 Introduction

Let $\mathcal{G}$ be an LTI-stable system with an unknown transfer function $G^0(p)$ defined as in Eq. (1.1)

$$y(t) = G^0(p)u(t) + v(t) \tag{1.1}$$

and $\mathbf{Y}^N$ be the given measurement data set as in Eq. (1.2). Let $G(\bar{p},\theta)$ be the estimated nominal transfer function. In terms of $G(\bar{p},\theta)$, the total modeling error is

$$e(k) = y(k) - G(\bar{p},\theta)u(k) = G_\Delta(\bar{p})u(k) + v(k), \tag{5.1}$$

where

$$G_\Delta(\bar{p}) = G^0(\bar{p}) - G(\bar{p},\theta)$$

is the error in transfer function estimation. Although $G_\Delta(\bar{p})$ is also a function of θ, it has been omitted here without loss of mathematical tractability. The effects of sampling on errors are neglected here.

Let $\{\epsilon(k)\} \in \varepsilon$. Given ε, error quantification involves computing a closed ball $\mathcal{S}(j\omega) \subset \mathcal{C}$ with some metric defined in $\mathcal{C}$, where $\mathcal{S}(j\omega)$ is distributed according to some $\mathcal{D}$ such that

$$\bar{G}(j\omega) = \begin{bmatrix} \Re\{G_\Delta(j\omega)\} \\ \Im\{G_\Delta(j\omega)\} \end{bmatrix} \in \mathcal{S}(j\omega), \forall\, \omega. \tag{5.2}$$

Here the nature of $\mathcal{D}$ depends on the prior ε. In conventional stochastic estimation when $\{\epsilon(k)\}$ is assumed to be normally distributed with known/unknown mean and variance, $\mathcal{D}$ denotes Chi-square distribution in two degrees of freedom (signifying the complex plane) and $\mathcal{S}$ is characterized by a closed ellipse given by

$$\mathbf{P}\left\{\bar{\mathbf{G}}(j\omega)^{\mathrm{T}}\mathbf{P}(j\omega)^{-1}\bar{\mathbf{G}}(j\omega) \leq \chi^2_{1-\alpha}(2)\right\} = \alpha,$$

where $\mathbf{P}(j\omega)$ is a positive definite matrix. Here the error is quantified with α-level confidence ellipses (Goodwin et al. 1992) at a given frequency ω. For each confidence level α, an ellipse is obtained with its size growing with increase in α.

On the other hand, in the case of SMI-based deterministic error quantification approaches, $\{e(k)\}$ is assumed to be uniformly distributed with known/unknown bounds [such as given in Eq. (4.31)]. Then $\mathcal{D}$ denotes uniform distribution with $\mathcal{S}(j\omega)$ characterized as a circle, an ellipse, or a rectangle, depending on the chosen metric. [See Norton (1987a) for an interesting discussion on parameter error bounding with ellipsoids, polytopes, etc.]

The major issues before the task of error quantification are the following:

1. Characterization of the unknown modeling error $\{e(k)\}$ such that $\{e(k)\} \in \varepsilon$. Note that $\{e(k)\}$ has two components, one due to model simplification and the other due to unmeasurable disturbances. This characterization, either in a stochastic framework or in a deterministic framework, is neither simple nor direct because both the components cannot in general be described together by the same ε.
2. Given ε, computation of $\mathcal{S}$. This is an inference problem where a realistic choice of ε is crucial for computation of accurate and valid $\mathcal{S}$.

The above may be tackled in two ways depending on the available prior information and its nature.

1. Either ε or sufficient information to obtain ε is known *a priori*. Then both the nominal transfer function and the associated $\mathcal{S}$ are estimated together from the measurement data. This is the well-known SMI approach to error quantification. See Wahlberg and Ljung (1992) and the references therein.
2. Only the nature of ε is known and whatever is known *a priori* is not sufficient to compute ε. Additional information has to be obtained from the measurement data alone. In this case, parameter estimation and error quantification cannot be made together. This is the case considered in the present chapter.

Thus, prior information plays a dominant role in the formulation of an error quantification algorithm.

5.1.1 Role of Priors

The problem of finding bounds on errors, as stated above, is a scientific inference problem. Prior assumptions as probability distributions of the unknown quantities (parameters, errors, etc.), which are supposed to represent what is known about the unknown plant before the actual data is available, play an important role in any inference problem. Such assumptions represent prior knowledge as well as relative ignorance of the observer. However, because data is expected to provide more than what is assumed/known *a priori*, and therefore to let the data speak for themselves, it is usually appropriate to consider the priors to be noninformative (Box and Tiao 1973). Though it is difficult to draw a precise line between informative and noninformative priors, we may intuitively say that a noninformative prior is one which lets the data speak for themselves, whereas an informative prior, by virtue of its large information content at an assumed higher confidence level, may make the inferential methodology (i.e., the identification algorithm) insensitive to the information contained in the data.

For example, a bound on $e(k)$ as

$$e(k) \leq r_k^2,\ r_k \text{ known,}$$

is an informative prior. Any quantification algorithm may altogether reject the data when the above priors are violated. On the other hand, a prior of the form

$$e(k) \leq r_k^2,\ r_k \text{ finite,}$$

is noninformative.

Two types of prior distributions of the unknown parameters are generally popular: Gaussian distribution, in a classical stochastic framework; and uniform distribution, in a deterministic robust framework. Leaving the debate on the choice among these two to Goodwin et al. (1992) and Wahlberg and Ljung (1992), because nearly all robust control design methods require explicit deterministic bounds on the modeling error of the existing plant, we follow a deterministic framework of error quantification.

Many suggestions have so far been made in the literature regarding the priors on modeling errors. For instance,

- Wahlberg and Ljung (1992) suggest the following prior on tail contribution, that is,

$$\sum_{k=n+1}^{\infty} \left|\delta_k\right| \leq \epsilon_k, \tag{5.3}$$

where $\{\delta_k\}$ is the impulse response sequence of $G_\Delta(q)$. Such a prior when combined with bounds on the input signal leads to bounds on $\{e(k)\}$, which in turn are mapped into the parameter space by the identification algorithm. However, it is difficult to realize such priors in practice.

- The following bound on multiplicative nonparametric uncertainty is suggested by Kosut et al. (1992):

$$\|\Delta_G\|_\infty \leq 1,$$

where

$$G_\Delta(q) = G(q, \theta)W_G(q)\Delta_G(q)$$

and $W_G(q)$ is a known frequency weighting function. Similar suggestion has also been made by van den Boom et al. (1991). It is generally said that "extensive prior analysis" of the measured input–output data should provide such bounds. An incorrect or overconservative choice of $W_G(q)$ may produce unbounded results as pointed out by Kosut et al. (1992).

- Another suggestion is

$$|\delta_k| \leq M\rho^{-k}, \forall k,$$

where M and ρ are known (Gu and Khargonekar 1992; Hakvroot 1993).

Such priors are essentially quantitative and considering the above discussion are informative. Moreover, these are difficult to realize in practice.

An exception to the above is that suggested by Goodwin et al. (1992):

$\{\delta_k\}$ is a Gaussian and independent process with

$$\mathcal{E}\{\delta_k^2\} = \alpha\lambda^k, \forall k,$$

where α and λ are "unknowns."

By modeling the error in this form, a model set parameterization is obtained that is described by a small number of parameters, yet, is capable of representing a large set of error models. However, the resulting bounds are not worst-case.

5.1.2 A Plausible Philosophy

At this end, we stress that measured data is the only available form of truth and priors only represent the observer's faith on the process being observed. Priors should, of course, be treated with skepticism. To let the data and not the observer speak of themselves (data) and because data can do so via a model only, whatever needs to be quantified is to be modeled. Such a model is then used to compute bounds on the modeling error. By this, we mean the following scheme:

- Obtain a nominal model of specified lower order.
- Compute residual sequence $\{e(k)\}$ and then fit an adequately high-order model to this.
- Then compute bounds on residuals and, consequently, on the nominal transfer function.

The priors involved here are:

- $\{v(k)\}$ is uniformly distributed.
- $G_{\Delta}(q)$ can be represented sufficiently accurately by an FIR model that is noninformative.

This seems to be the only possible way of quantifying the modeling error when we are constrained with noninformative priors. Such a scheme has some beneficial features:

- The nominal model may be chosen/estimated to satisfy any control relevant criterion such as bias distribution, then the above scheme provides worst-case error bounds concerning such a nominal model.
- Obtained bounds will be realistic and as such the measured data cannot provide smaller bounds.
- In the case of iterative schemes of combined identification and control design (Gevers 1993), one need not worry about what priors to choose (an obvious test of faith!) in each iteration given that this scheme is less sensitive to the chosen noninformative priors.

The only approach that comes under the above philosophy appears to be that of Goodwin et al. (1992). Here, though in a completely stochastic framework under Gaussian assumptions, nominal model and as well as error bounds are estimated from the measured data themselves, not making use of any informative prior. There are criticisms, for example, Kosut (1993) and Gevers (1991), that estimating the error as a parameterized model is tantamount to estimating a higher-order model that should have been incorporated in the nominal plant model. However, *parameterization* *of G_Δ as a higher order model only to compute error bounds* does not amount to *estimating modeling error.* Moreover, this is unavoidable when the available priors are at question.

5.1.3 Chapter Layout

On the above lines, a simple off-line error quantification scheme is suggested in this chapter. Given a finite-length measurement data record, this scheme delivers ellipsoidal bounds on parameter errors. For computation of error bounds, the only assumption made is that G_Δ has a finite impulse response. Both the nominal model and the associated error bounds are computed from the same measurement data record, without making use of any informative prior. The computed bounds are, however, conservative. The aim of this chapter is only to show that it is still possible to compute bounds quantifying the modeling error even without informative priors.

The layout of the rest of the chapter is as follows: The robust parameter estimation problem under an $\mathcal{H}_\infty$ criterion is stated in Section 5.2. In Section 5.3 the proposed error quantification scheme is derived. Numerical simulation examples are presented in Section 5.4, and concluding remarks are made in Section 5.5.

5.2 Robust Parameter Estimation

To estimate a model of the system $G^0(p)$, consider the input–output model set $\mathcal{M}$ defined as

$$\mathcal{M} := \left\{ y = G(p,\theta)u + e : G(p,\theta) \in \mathcal{M}^*_{\text{prior}}, e \in \varepsilon \right\}, \tag{5.4}$$

where $\mathcal{M}^*_{\text{prior}}$ is the prior set of transfer functions

$$\mathcal{M}^*_{\text{prior}} = \left\{ G(\bar{p},\theta) = \sum_{i=1}^{d} \mathcal{F}_i(\bar{p})\theta_i ; \theta = [\theta_1, \theta_2, \cdots, \theta_d]^{\mathrm{T}} \in \mathcal{S}^d(\hat{\theta}(0), \mathbf{P}_\theta(0)) \right\} \tag{5.5}$$

and

$$\varepsilon = \left\{ e : \|e\|_2^2 \le C_e ; C_e \in \mathbb{R}^+ \right\}. \tag{5.6}$$

$\{\mathcal{F}_i(q)\}$ is a set of linear stable filters forming a basis for the model set $\mathcal{M}^*$. $\mathcal{S}^d(\hat{\theta}(0), \mathbf{P}_\theta(0) \subset \mathbb{R}^d$ is a closed ellipsoid defined as

$$\left\{ \theta : \left[\theta - \hat{\theta}(0)\right]^{\mathrm{T}} \mathbf{P}_\theta(0)^{-1} \left[\theta - \hat{\theta}(0)\right] \le 1 \right\}. \tag{5.7}$$

Here $\hat{\theta}(0)$ is the prior estimate of θ, and $\mathbf{P}(0)$ is a positive definite matrix. The set $\mathcal{S}^d(\hat{\theta}(0), \mathbf{P}_\theta(0))$ is the prior parameter set in the SMI parlance. Given $\mathbf{Y}^N$, the problem is to find the posterior set of models $\mathcal{M}^*_{\text{post}}$,

$$\mathcal{M}^*_{\text{post}} = \mathcal{M}^*_{\text{prior}} \cap \left\{\theta : y(k) = \Phi^{\mathrm{T}}(\overline{p})\theta + e(k), e \in \varepsilon\right\}, \tag{5.8}$$

where

$$\Phi(k) = \left[\mathcal{F}_1(\overline{p})u(k), \cdots, \mathcal{F}_d(\overline{p})u(k)\right]^{\mathrm{T}} \tag{5.9}$$

$$\theta = \left[\theta_1, \cdots, \theta_d\right]^{\mathrm{T}} \tag{5.10}$$

5.3 Quantification of Parameter Errors

Given prior sets $\mathcal{M}^*_{\text{prior}}$ and ε along with $\mathbf{Y}^N$, let there exist an estimator Γ with $\|T\|_\infty < \gamma, \gamma$ finite (in the case of NBLS) or infinite (in the case of LS), yielding an estimate $\hat{\theta}^{(N)}$. Then the posterior set of transfer functions quantifying the estimation errors is given by, as shown in the previous chapter,

$$\mathcal{M}^*_{\text{post}} = \left\{G(q, \theta) : \theta \in \mathcal{S}^d\left(\hat{\theta}(N), \mathbf{P}(N)\right) \in \mathbb{R}^d\right\}, \tag{5.11}$$

where

$$\mathbf{P}(N) = \left[\mathbf{P}(0)^{-1} + \sum_{k=1}^{N} \Phi(k)\Phi(k)^{\mathrm{T}}\right]^{-1} \sigma(N) \tag{5.12}$$

$$\sigma(N) = 1 + C_e - \sum_{k=1}^{N} \epsilon(k)^2 - \left[\hat{\theta}(N) - \hat{\theta}(0)\right]^{\mathrm{T}} \mathbf{P}(0)^{-1} \left[\hat{\theta}(N) - \hat{\theta}(0)\right] \tag{5.13}$$

and

$$\epsilon(k) = y(k) - \Phi^{\mathrm{T}}(k)\hat{\theta}(N). \tag{5.14}$$

Recall that

$$C_e = \sum_{k=1}^{N} r_k, r_k \ge e(k)^2.$$

From the above set of equations it is clear that, when $\hat{\theta}(N)$ and $\bar{\mathbf{P}}(N)$ are available after parameter estimation, the size of the ellipsoid $\mathcal{S}^d$ is governed by $\sigma(N)$ alone. However, because $\sigma(N)$ involves the unknown C_e, accuracy and tightness of $\mathcal{S}^d$ depend on the accurate and valid choice of C_e. Any over-conservative choice of C_e will lead to a large $\sigma(N)$, and the estimated model may show poor performance in any application such as robust control design where such bounds are used for design. In the present framework of error quantification, the crux of the problem thus lies in the computation of C_e from the measurement data.

We now set out to compute C_e, from the measured data themselves, along with the nominal model $G(\bar{p},\hat{\theta}(N))$ in an off-line setting. By separating the two phases of nominal model estimation and computation of the associated error bounds in this manner, we can impose any application-relevant weighting function in the former.

Given $G(\bar{p},\hat{\theta}(N))$, let

$$e(k) = y(k) - G(\bar{p},\hat{\theta}(N))u(k) \tag{5.15}$$

$$= G_\Delta(q)u(k) + v(k). \tag{5.16}$$

We now assume that the unmodeled part of $G(q)$, that is, $G_\Delta(q)$ has a finite impulse response, that is,

$$G_\Delta(q) = \sum_{i=1}^{l} \eta_i q^{-i}, l \le N. \tag{5.17}$$

Under this assumption, we can model $e(k)$ as

$$e(k) = \psi^{\mathrm{T}}\eta + v(k), \tag{5.18}$$

where $\eta = [\eta_1, \cdots, \eta_l]^{\mathrm{T}}$ and $\psi = [u(k-1), \cdots, u(k-l)]^{\mathrm{T}}$.

Given $\mathbf{Y}^N$ and $G(\bar{p},\hat{\theta}(N))$, let $\hat{\eta}(N)$ be the NBLS estimate of η such that

$$e(k) = \sum_{i=1}^{l} \hat{\eta}_i q^{-i} u(k) + \hat{v}(k), \tag{5.19}$$

where $\{\hat{v}(k)\}$ is the computed residual sequence. Under the assumption of uniform distribution of the random disturbances, assume that

$$|v|_\infty^2 \le C_v, \tag{5.20}$$

where

$$C_v = C_v' \left|\hat{v}\right|_\infty^2, C_v' > 1.$$

Then, we can compute the set

$$\left\{\eta : \eta \in \mathcal{S}^l\left(\hat{\eta}(N), \mathbf{P}_\eta(N)\right) \subset \mathbb{R}^l\right\}, \tag{5.21}$$

where

$$\mathbf{P}_\eta(N) = \left[\bar{\mathbf{P}}_\eta(N)^{-1} + \gamma^2 \sum_{k=1}^{N-1} \psi(k)\psi(k)^\mathrm{T}\right]^{-1} \sigma_\eta(N), \tag{5.22}$$

$$\sigma(N) = 1 + NC_v - \left\|\hat{v}\right\|_2^2, \tag{5.23}$$

$$\bar{\mathbf{P}}_\eta(N) = \sum_{k=1}^{N} \psi(k)\psi(k)^\mathrm{T} - \gamma^2 \sum_{k=1}^{N-1} \psi(k)\psi(k)^\mathrm{T} > 0. \tag{5.24}$$

The ellipsoid $\mathcal{S}^l\left(\hat{\theta}_\eta(N), \mathbf{P}_\eta(N)\right)$ now describes the unmodeled dynamics G_Δ. By computing the distance between the center $\hat{\theta}_\eta(N)$ of this ellipsoid and a hyperplane tangential to the ellipsoid and normal to $\psi(k)$, the effect of G_Δ on y can be quantified. Thus,

$$e(k) \le r_k^{1/2} = \frac{\psi^\mathrm{T}(k)\mathbf{P}_\eta(k)\psi(k)}{\psi(k)^\mathrm{T}\psi(k)} + \mathrm{abs}\left[\psi^\mathrm{T}(k)\hat{\eta}(N)\right] + C_v^{1/2}. \tag{5.25}$$

For proof see Norton (1987a, 1987b). Thus,

$$C_e = \sum_{k=1}^{N} \left\{\frac{\psi^\mathrm{T}(k)\mathbf{P}_\eta(k)\psi(k)}{\psi(k)^\mathrm{T}\psi(k)} + \mathrm{abs}\left[\psi^\mathrm{T}(k)\hat{\eta}(N)\right]\right\} + NC_v, \tag{5.26}$$

Using Eqs. (5.11) through (5.14), we finally obtain

$$\mathcal{M}_\mathrm{post}^* = \left\{G(\bar{p}, \theta) : \theta \in \mathcal{S}^d\left(\hat{\theta}(N), \mathbf{P}_\theta(N)\right) \in \mathbb{R}^d\right\}. \tag{5.27}$$

This set describes an ellipsoid in $\mathbb{R}^d$. Instead, the same may be interpreted as an ellipse in $\mathcal{C}$ using the following result of Wahlberg and Ljung (1992).

The set

$$\left\{\theta : \left[\theta - \hat{\theta}(N)\right]^\mathrm{T} \mathbf{P}_\theta(N)^{-1} \left[\theta - \hat{\theta}(N)\right] \le 1\right\} \tag{5.28}$$

ensures that

$$\bar{G}(j\omega)^{T}\mathbf{P}_{G}(j\omega)^{-1}\bar{\mathbf{G}}(j\omega)\leq 1, \tag{5.29}$$

$$\mathbf{P}_{G}(j\omega)=\begin{bmatrix}\Re\{\mathcal{B}(j\omega)\}\\ \Im\{\mathcal{B}(j\omega)\}\end{bmatrix}\mathbf{P}_{\theta}(N)\left[\Re\{\mathcal{B}(j\omega)\} \quad \Im\{\mathcal{B}(j\omega)\}\right], \tag{5.30}$$

where $\bar{\mathbf{G}}(j\omega)$ is as defined in Eq. (5.2) and $\mathcal{B}(j\omega)$ is the basis evaluated in the complex plane at a complex frequency $j\omega$. Thus, an ellipse is obtained at each ω, quantifying the error.

5.4 Illustrative Examples

Example 5.1

Consider a continuous system:

$$G^{0}(s)=\frac{2s+6}{s^{3}+7s^{2}+11s+6}$$

This is simulated with a uniform random signal (2.0 ± 2.0) input for 30 s at a sampling time 0.1 s such that $N = 300$. A similar signal of amplitude ± 0.1 uncorrelated with the input signal is added as disturbance to the simulated output signal.

A second-order model using basis BSVF is estimated with $\lambda = 1$, using the RLS algorithm, as

$$G(p,\hat{\theta})=\frac{0.3168p+1.0081}{(p+1)^{2}}$$

with $P_{\theta}(0)=10^{4}\mathbf{I}$.

The following procedure is followed for error quantification.

- G_{Δ} is estimated as a 30th-order ($l = 30$) FIR model [Eq. (5.18)].
- C_{v} is computed as a bound on the residuals [Eq. (5.20)]. We obtain $C_{v} = 0.0233$. Note that this is a conservative estimate of the actual bound (0.01) on the disturbance term.
- The ellipsoid of Eq. (5.21) is computed from Eqs. (5.22) through (5.24).
- C_{e} is computed from Eq. (5.26).
- Parameter error bounding ellipse $\mathcal{S}^{2}\left(\hat{\theta}(300),\mathbf{P}_{\theta}(300)\right)$ is computed from Eqs. (5.12) through (5.14). We obtain

$$\mathbf{P}(300)=\begin{bmatrix}0.00832 & -0.00808\\ -0.00808 & 0.47965\end{bmatrix}.$$

- Finally, frequency domain ellipses as in Eq. (5.29) are computed using Eq. (5.30).

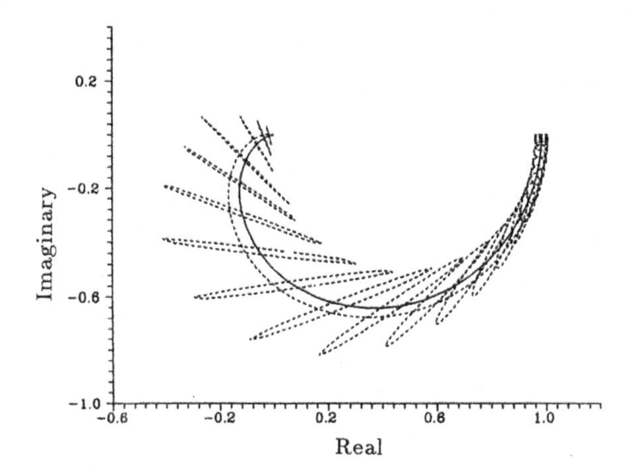

FIGURE 5.1
Frequency domain error bounds (Example 5.1) (dotted line, true system; continuous line, estimate).

Figure 5.1 shows the frequency domain elliptical bounds on the transfer function estimation error. Here the dotted and the continuous curves are the Nyquist plots of the transfer functions of the true system and the estimate respectively. The ellipses are drawn with their centers lying on the Nyquist plot of the estimate at various frequencies. Thus, each ellipse represents a set of estimates ($G(j\omega,\hat{\theta})$ being a member of the set) quantifying the error at a particular frequency ω. Due to the low-pass nature of the chosen model structure, a good low-frequency fit is obtained. The same is reflected in the bounding ellipses also.

Example 5.2

Reconsider Example 4.1. The system is simulated with the same set of input and disturbance signals as that of Example 5.1, but for $N = 600$.

Using the NBLS algorithm, the following model is estimated at $\lambda = 1.145$.

$$G(p,\hat{\theta}) = \frac{-0.04844p + 24.5877}{(p+5)^2}.$$

By constructing a 30th-order FIR model for G_Δ, and forming the residuals, C_v is estimated to be 0.744. Note that this is a very crude estimate of noise characteristics. Using a similar procedure as detailed for the previous example, we finally obtain

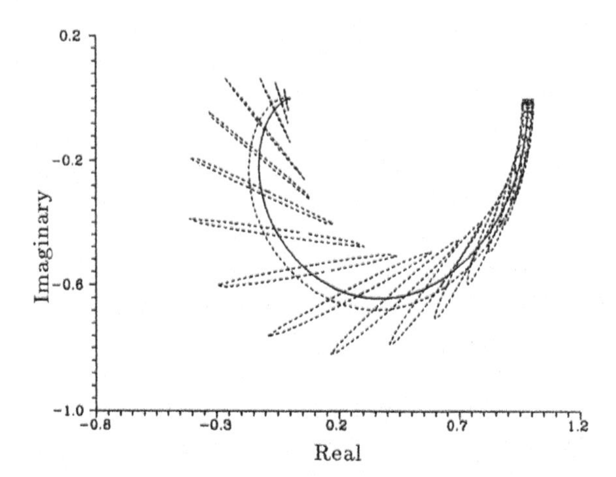

FIGURE 5.2
Frequency domain error bounds (Example 2.2) (dotted line, true system; continuous line, estimate).

$$\mathbf{P}(600) = \begin{bmatrix} 0.37016 & -0.00356 \\ -0.00356 & 0.61004 \end{bmatrix}$$

and the corresponding frequency domain elliptical bounds are shown in Figure 5.2.

5.5 Conclusions

The observation that the true system is contained in the ellipses bounding the nominal model estimate validates the suggested error quantification scheme (Figures 5.1 and 5.2). However, because both the nominal model and the associated bounds are computed from the same measurement data record, and because the priors are noninformative, the computed bounds are conservative. Note that conservatism is always associated with robustness. With the kind of priors assumed here, no tighter bounds can be derived. Only by increasing the information content in the priors it is possible to get tighter bounds, for which novel techniques are already available in the literature.

In the literature on SMI and error quantification, the size of the posterior set of models is considered as a measure of performance of the algorithm used for quantification. However, such a measure is questionable. By considering information-rich priors on modeling errors, it is always possible (provided such priors are valid) to come up with a good performance index. A more general performance criterion should weigh sizes of both priors on modeling errors and posterior parameter sets. The performance of the proposed scheme should be judged in this perspective.

6

Conclusions

In this book, some aspects of linear and robust estimation of CT models from sampled measurement data have been investigated. Although some of the results are encouraging, the others point to some open issues for further investigations. These are discussed in the following.

6.1 Linear Model Parameterizations for System Identification

One of the crucial phases of parametric system identification is the parameterization of the model structure. Traditionally, models are parameterized as rational transfer functions. Such descriptions, despite their clarity of dynamics with poles and zeros, have some limitations in the context of system identification. A limitation of concern is that such parameterizations being inherently nonlinear lead to nonlinear estimation. Attempts to transform them into linear parameterizations are, however, not trouble free. These have been pointed out in Chapter 1. It has also been pointed out that linearly parameterized model structures simplify the problem, with linear estimation being the main advantage.

In a class of linear model parameterizations, CT models parameterized with MPs and TMs have been studied in this book for irreducible model identification of MIMO systems. An advantage with these parameterizations is that irreducible rational transfer function (matrix) or minimal state–space descriptions of models are realizable even from truncated sequences of MPs and TMs. Thus, the sense of poles and zeros is preserved, which is of importance in other branches of control. For a limited class of finite dimensional systems for which such zeros and poles do exist, this book shows that it is possible to capture them in an iterative fashion. The proposed iterative scheme has been analyzed for its convergence.

The definitions of MP and TM models have been generalized and made flexible. The flexibility comes from the fact that prior knowledge of system modes can be effectively used to choose the coefficients λ and β used in these models. In particular, MP models have been so generalized as to handle even resonant systems as well. When more information about the approximate location of poles of the system is available, it has been shown that, by embedding such information in the basis as poles, it is possible to get good approximations with a small number of parameters.

Due to the linearity of the suggested parameterizations, the estimation is robust to zero-mean white/colored disturbances. The Monte Carlo simulation results of Chapters 2 and 3 confirm this. Although "unbiased" rather than "robust" estimation may appear to be the correct word to denote this feature, the latter is preferred for the following reason. An estimate $G(j\omega,\hat{\theta})$ is said to be unbiased if $E\{G(j\omega,\hat{\theta})\} \rightarrow G^0(j\omega)$, where $E\{.\}$ denotes the expectation operator. Due to inherent undermodeling, unbiased estimates can never be obtained with finite dimensional models.

The problem of inflated models for MIMO systems has been avoided because the chosen parameterizations do not involve unknown denominators. Recall that in the case of rational TFM models with EE minimization, formation of a common denominator inflates the size of the unknown parameter vector.

6.2 Robust Estimation

Considering the contributions by several researchers in the field of filtering under an $\mathcal{H}_\infty$ performance criterion, the problem of parameter estimation has been studied in Chapter 4 under a similar performance criterion. The proposed $\mathcal{H}_\infty$-NBLS algorithm is a special case of the set of $\mathcal{H}_\infty$ filter equations. By definition, such a criterion minimizes the estimation error energy over all possible modeling errors of fixed energy. In the context of parameter estimation this has some interesting consequences. The NBLS estimator makes cautious information updates leading to active and accelerated estimation in the case of finite measurement data. However, the response of the NBLS algorithm is different from that of the conventional RLS algorithm only in the non-asymptotic case. Asymptotically, as the number of measurements tend to infinity, the NBLS estimate approaches the RLS estimate.

In the fields of parameter estimation and adaptive filtering, there do exist numerous algorithms for cautious estimation with the ability to track time-varying parameters. Examples are RLS algorithms with a variable forgetting factor and variants of least-mean square and other gradient-type adaptive estimation algorithms. The NBLS algorithm has been observed to achieve these in a different way. However, further experimentation and analysis are required toward comparing the performance and computational requirements of the NBLS algorithm with other existing algorithms with similar features. In this connection, note that the NBLS algorithm is computationally more expensive than that of the simple RLS algorithm.

In the case of $\mathcal{H}_\infty$ filtering, a common practice is to assume that the $\mathcal{H}_\infty$ norm of the filter is known *a priori*. Instead, an explicit formula has been suggested in this book to compute such a norm from the available measurement records such that convergence of the estimation algorithm is guaranteed.

To make the NBLS algorithm suitable for SMI, non-asymptotic formulae have been derived to compute deterministic ellipsoidal parameter error bounds.

6.3 Error Quantification

Error quantification is important to ascertain model quality and thereby render the identification effort complete. In recent years, there has been a hectic activity in the field of system identification toward error quantification. This issue has been studied in Chapter 5.

A survey of some of the existing literature on error quantification has been made to reveal the nature of prior information used. It has been noticed that most of the existing approaches to error quantification require information-rich priors. However, in reality, it is difficult to realize such priors. The need for quantification approaches that do not require informative priors has been stressed while giving some ideas on what such an approach should look like.

To illustrate these ideas, a simple off-line scheme has been suggested that quantifies error with deterministic bounds, making use of noninformative priors. In this scheme, a robust estimate (both nominal model and the associated membership set) is computed from the given measurement record alone. The conservative estimates are a characteristic of robust estimation.

In the proposed setup for error quantification, only a limited class of LTI systems has been considered. Here the modeling error is due to reduced-order model estimation along with the contribution of disturbances. However, when the system under test is time-varying or when we are interested in the estimation of linearized models of nonlinear systems over a specified change in operating points, the nature of noninformative priors considered in this book will lead to very crude error bounds. Such results may not serve the purpose for which the exercise of error quantification is conducted. Further investigations are to be made in these directions.

To conclude, the problem of linear and robust estimation of CT models of LTI CT systems has been addressed in this book. The suggested linear parameterizations lead to linear estimation that is asymptotically robust to zero-mean additive disturbances. To quantify the total modeling error in the estimates, the suggested scheme delivers robust estimates making use of easily realizable noninformative priors.

Bibliography

Anderson, B.D.O.: Identification of scalar errors-in-variables models with dynamics, *Automatica*, 1985, 21 (6), pp. 709–716.

Anderson, B.D.O., Moore, J.B., and Hawkes, R.M.: Model approximations via prediction error identification, *Automatica*, 1978, 14 (9), pp. 615, 622.

Arruda, L.V.R., and Favier, G.: A review and a comparison of robust estimation methods. *Proceedings of IFAC Symposium on Identification and System Parameter Estimation*, Budapest, Hungary, 1991.

Aström, K.J., and Eykhoff, P.: System identification—A survey, *Automatica*, 1971, 7 (2), pp. 123–162.

Balchen, J.: How have we arrived at the present state of knowledge in process control? Is there a lesson to be learned? *J. Process Contr.*, 1999, 9, pp. 101–108.

Bapat, V.N.: Some extensions to Poisson moment functional based estimation of continuous-time models of dynamical systems. PhD Thesis, Indian Institute of Technology, Kharagpur, India, 1993.

Bastogne, T., Garnier, H., and Sibille, P.: A PMF-based subspace method for continuous-time model identification—Application to a multivariable winding process, *Int. J. Contr.*, 2001, 74 (2), pp. 118–132.

Bastogne, T., Garnier, H., Sibille, P., and Mensler, M.: Applicability of Poisson moment functionals to subspace methods for continuous time system identification. *Proceedings of the 11th IFAC Symposium on System Identification* (*SYSID97*), Fukuoka, Japan, 1997, 4, pp. 1665–1670.

Bierman, G.J.: *Factorization Methods for Discrete Sequential Estimation*, Academic Press, New York, 1977.

Billings, S.A.: Identification of nonlinear systems—A survey, *Proc. IEE D Contr. Theor. Appl.*, 1980, 127 (6), pp. 272–285.

Billings, S.A., and Lang, Z.: Truncation of nonlinear system expansions in the frequency domain, *Int. J. Contr.*, 1997, 68 (5), pp. 1019–1042.

Bohn, C., and Unbehauen, H.: Sensitivity models for nonlinear filters with application to recursive parameter estimation for nonlinear state-space models, *Proc. IEE. Contr. Theor. Appl.*, 2001, 148 (2), pp. 137–145.

Bohn, C., and Unbehauen, H.: The application of matrix differential calculus for derivation of simplified expressions in approximate nonlinear filter algorithms, *Automatica*, 2000, 36, pp. 1553–1560.

Box, G.E.P., and Tiao, G.C.: *Bayesian Inference in Statistical Analysis*, Addison-Wesley Publishing Company, Reading, MA, 1973.

Bultheel, A., and van Barel, M.: Pade techniques for model reduction in linear system theory: A survey, *J. Comput. Appl. Math.*, 1986, 14, pp. 401, 438.

Chakravarty, B., Mandyam, N., Mukhopadhyay, S., Patra, A., and Rao, G.P.: Real-time parameter estimation via block pulse functions. *Proceedings of SICE89*, Matsuyama, Japan, 1989, pp. 1095–1098.

Chen, C.F., and Hsiao, C.H.: Time domain synthesis via Walsh functions, *Proc. IEE*, 1975, 122 (5), pp. 565–570.

Clymer, A.B.: Direct system synthesis by means of computers, *Trans. Amer. Inst. Electr. Eng.*, 1959, 77 (I), pp. 798–806.

Co, T.B., and Ungarala, S.: Batch scheme recursive parameter estimation of continuous-time systems using the modulating function method, *Automatica*, 1997, 33 (6), pp. 1185–1191.

Coirault, P., Trigeassou, J.C., Gaubert, J.P., and Champenois, Q.: Parameter identification of an induction machine for diagnosis. *Proceedings of the IFAC-Safe Process, 97*, Kingston upon Hull, UK, 1997, pp. 276–281.

Corrington, M.S.: Solution of differential and integral equations with Walsh functions, *IEEE Trans. Circuit Theory*, 1973, 20 (5), pp. 470–476.

Daniel-Berhe, S.: *Parameter Identification of Nonlinear Continuous Time Systems Using the Hartley Modulating Functions Method*, Cuvillier-Verlag, Göttingen, Germany, 1999.

Daniel-Berhe, S., and Unbehauen, H.: Batch scheme recursive Hartley modulating functions identification of nonlinear continuous-time Hammerstein model. *Proceedings of the European Control Conference* (*ECC-99*), Karlsruhe, Germany, 1999, paper 1013-3.

Daniel-Berhe, S., and Unbehauen, H.: Bilinear continuous-time systems identification via Hartley-based modulating functions, *Automatica*, 1998, 34 (4), pp. 499–503.

Daniel-Berhe, S., and Unbehauen, H.: Identification of nonlinear continuous-time Hammerstein model via HMF-method. *Proceedings of the 36th IEEE Conference on Decision and Control* (CDC), San Diego, CA, 1997, pp. 2990–2995.

Daniel-Berhe, S., and Unbehauen, H.: Parameter estimation of nonlinear continuous-time systems using Hartley modulating functions. *Proceedings of the IEE UKACC International Conference on Control*, Exeter, UK, 1996, pp. 228–233.

Daniel-Berhe, S., and Unbehauen, H.: Physical parameter estimation of the nonlinear dynamics of a single link robotic manipulator with flexible joint using the HMF-method. *Proceedings of the 16th American Control Conference* (*ACC*), Albuquerque, NM, 1997, pp. 1504–1508.

Daniel-Berhe, S., and Unbehauen, H.: Physical parameters estimation of the nonlinear continuous-time dynamics of a DC motor using HMF-method, *J. Franklin Inst.*, 1999, 336, pp. 481–501.

Daniel-Berhe, S., and Unbehauen, H.: State space identification of bilinear continuous-time canonical systems via batch scheme Hartley modulating functions approach. *Proceedings of the 37th International Conference Decision Control* (*CDC*), Tampa, FL, 1998, pp. 4482–4487.

Dastych, J., and Kueper, P.: Identifikation zeitkontinuierlicher Systeme mit Block-Puls-Funktionen und bilinearen Transformationen, *Automatisierungstechnik*, 1995, 43 (1), pp. 24–30.

Datta, K.B., and Mohan, B.M.: *Orthogonal Functions in Systems and Control*, World Scientific, Singapore, 1994.

De Moor, B., Vandevalle, J., Vanderberghe, L., and Van Mieghem, P.: A geometrical strategy for identification of state space models of linear multivariable systems with singular value decomposition. *Proceedings of the IFAC-Symposium Identification and System Parameter Estimation*, Beijing, 1988, pp. 700–704.

Deller, J.R.: Set membership identification in digital signal processing. *IEEE ASSP Magazine*, 1989, 6 (4), p. 20.

Dhawan, R.K., Sahai, A., Nishar, D.V., and Rao, G.P.: Recursive estimation of Markov parameters in linear continuous-time SISO systems via block pulse functions. *Proceedings of the 9th IFAC/IFORS Symposium Identification and System Parameter Estimation*, Budapest, Hungary, 1991, 2, pp. 1495–1500.

Diamessis, J.E.: A new method of determining the parameters of physical systems, *Proc. IEEE*, 1965a, 53, pp. 205–206.

Diamessis, J.E.: On the determination of the parameters of certain nonlinear systems, *Proc. IEEE*, 1965b, 53, pp. 319–320.

Diamessis, J.E.: A new method of determining the parameters of certain time-varying systems, *Proc. IEEE*, 1965c, 53, pp. 396–397.

Diekmann, K., and Unbehauen, H.: Recursive identification of multiple-input multiple-output systems. *Proceedings of the 5th IFAC Symposium Identification and System Parameter Estimation*, Darmstadt, Germany, 1979, pp. 423–429.

Dorato, P.: *Robust Control*, IEEE Press, New York, 1987.

Dorato, P., and Yedawalli, R. (Eds.): *Recent Advances in Robust Control*, IEEE Press, New York, 1990.

Evans, C., Weiss, M., Escobet, T., Quevedo, J., Rees, D., and Jones, L.: Identification in the time and frequency domains: A comparison using the ECC bench-mark model. *Proceedings of the UKACC International Conference on Control*, University of Exeter, UK, 1966, pp. 1290–1296.

Eykhoff, P.: *System Identification*, Wiley, New York, 1974.

Fairman, F.W.: Parameter identification for a class of multivariable nonlinear processes, *Int. J. Syst. Sci.*, 1971, 1 (3), pp. 291–296.

Fairman, F.W., and Shen, D.W.C.: Parameter identification for linear time-varying dynamic processes, *Proc. IEE*, 1970, 117, pp. 2025–2029.

Fletcher, R.: *Practical Methods of Optimization*, Wiley, New York, 1991.

Fogel, E., and Huang, Y.F.: On the value of information in system identification—Bounded noise case, *Automatica*, 1982, 18 (2), pp. 229, 238.

Gantmacher, F.R.: *Applications of the Theory of Matrices*, Interscience Publishers, New York, 1959.

Garnier, H., Gilson, M., and Zheng, W.X.: A bias-eliminated least squares method for continuous-time model identification of closed loop systems, *Int. J. Contr.*, 2000, 73 (1), pp. 38–48.

Garnier, H., and Mensler, M.: Comparison of sixteen continuous time system identification methods with the CONTSID toolbox. *Proceedings of the European Control Conference ECC99*, Karlsruhe, Germany, paper no. dm 3-6, 1999.

Garnier, H., Mensler, M., and Richard, A.: Continuous-time model identification from sampled data. Implementation issues and performance evaluation, *Int. J. Contr.*, 2003, 76 (13), pp. 1337–1357.

Garnier, H., Sibille, P., and Bastogne, T.: A bias-free least-squares parameter estimator for continuous-time state-space models. *Proceedings of the 36th IEEE Conference on Decision and Control* (*CDC97*), San Diego, CA, 1997, pp. 1860–1865.

Garnier, H., Sibille, P., and Nguyen, H.L.: A new bias-compensating least-squares method for continuous-time MIMO system identification applied to a laboratory-scale process. *Proceedings of the 3rd IEEE Conference on Control Application* (*CCA94*), Glasgow, Scotland, 1994, pp. 1711–1716.

Garnier, H., Sibille, P., and Richard, A.: Continuous-time canonical state-space model identification via Poisson moment functionals. *Proceedings of the 34th IEEE Conference on Decision Control* (CDC95), New Orleans, LA, 1995, pp. 3004–3009.

Garnier, H., Sibille, P., and Richard, A.: Recursive least-squares estimation of continuous models via generalised Poisson moment functionals. *Proceedings of the VII European Signal Processing Conference* (EUSIPCO94), Edinburgh, Scotland, 1994a, pp. 518–521.

Garnier, H., Sibille, P., and Spott, T.: Influence of the initial covariance matrix on recursive LS estimation of continuous models via generalised Poisson moment functionals. *Proceedings of the 10th IFAC Symposium on System Identification* (*SYSID94*), Copenhagen, Denmark, 1994b, pp. 3669–3674.

Garnier, H., Sibille, P., Mensler, M., and Richard, A.: Pilot crane identification and control in presence of friction. *Proceedings of the 13th IFAC World Congress*, San Francisco, CA, 1996, pp. 107–112.

Garnier, H., Sibille, P., Nguyen, H.L., and Spott, T.: A bias compensating least-squares method for continuous-time system identification via Poisson moment functionals. *Proceedings of the 10th IFAC Symposium on System Identification* (*SYSID94*), Copenhagen, Denmark, 1994c, pp. 3675–3680.

Garnier, H., Söderström, T., and Yuz, J.I. (Eds.): Special issue on continuous-time model identification, *IET Contr. Theor. Appl.*, 2011, 5 (7), pp. 839–942.

Garnier, H., and Wang, L. (Eds.): *Identification of Continuous-time Models from Sampled Data*, Springer, London, UK, 2008.

Garnier, H., and Young, P. (Eds.): Special issue on applications of continuous-time model identification and estimation, *Int. J. Contr.*, 2014, 87 (7), pp 1317–1466.

Gawthrop, P.J.: *Continuous-time Self-tuning Control*. IEE Publication Series, Research Studies Press, Letchworth, UK, 1987.

Gevers, M.: Connecting identification and robust control: A new challenge. *Proceedings of IFAC Symposium on Identification and System Parameter Estimation*, Budapest, Hungary, 1991.

Gevers, M.: Essays on control: Perspectives in the theory and its applications, chapter *Towards a Joint Design of Identification and Control*, pp. 111, 151. Volume 14 of Progress in Systems and Control Theory, Birkhäuser, Boston, MA, 1993.

Goodwin, G.C., and Sin, K.S.: *Adaptive Filtering, Prediction and Control*. Prentice Hall, Englewood Cliffs, NJ, 1984.

Goodwin, G.C., Gevers, M., and Ninness, B.: Quantifying the error in estimated transfer functions with application to model order reduction, *IEEE Trans. Automat. Contr.* 1992, 37 (7), pp. 913, 928.

Goodwin, G.C., Gevers, M., and Mayne, D.Q.: Bias and variance distribution in transfer function estimation. *Proceedings of IFAC Symposium on Identification and System Parameter Estimation*, pp. 952, 957, Budapest, Hungary, 1991b.

Goodwin, G.C., and Middleton, R.H.: *Digital Control and Estimation: A Unified Approach*. Prentice Hall, Upper Saddle River, NJ, 1990.

Goodwin, G.C., Ninness, B., and Poor, V.: Choice of basis functions for continuous and discrete system modelling. *Proceedings of the 9th IFAC Symposium on Identification and System Parameter Estimation*, Budapest, Hungary, 1991, pp. 1179–1184.

Greblicki, W.: Continuous-time Wiener system identification, *IEEE Trans. Automat. Contr.*, 1998, 43 (10), pp. 1488–1492.

Gu, G., and Khargonekar, P.P.: Linear and nonlinear algorithms for identification in H with error bounds, *IEEE Trans. Automat. Contr.*, 1992, 37 (7), pp. 953, 963.

Guillaume, P., Pintelon, R., and Schoukens, J.: Robust parametric transfer function estimation using complex logarithmic frequency response data. *Proceedings of the 10th IFAC Symposium on System Identification*, Copenhagen, Denmark, 1994, 2, pp. 495–502.

Haber, R., and Keviczky, L.: *Nonlinear System Identification-Input-Output Modelling Approach*. Vol. 1 & 2, Kluwer, Dordrecht, the Netherlands, 1999.

Hachimo, T., Karube, I., Minari, Y., and Takata, H.: Continuous time identification of nonlinear systems using radial basis function network model and genetic algorithm. *Proceedings of the 12th IFAC Symposium on System Identification Parameter Estimation*, Santa Barbara, CA, 2000, paper 141.

Hakvroot, R.G.: Worst-case system identification in $\mathcal{H}_\infty$: error bounds and optimal bounds. In *Preprints of the IFAC World Congress*, Sydney, Australia, 1993.

Hassibi, B., Sayed, A.H., and Kailath, T.: LMS is $\mathcal{H}_\infty$ optimal. *Proceedings of IEEE Conference on Decision and Control*, San Antonio, TX, 1993a.

Hassibi, B., Sayed, A.H., and Kailath, T.: Recursive linear estimation in Krein spaces—Part I: Theory. *Proceedings of IEEE Conference on Decision and Control*, San Antonio, TX, 1993b.

Hassibi, B., Sayed, A.H., and Kailath, T.: Recursive linear estimation in Krein spaces—Part II: Applications. *Proceedings of IEEE Conference on Decision and Control*, San Antonio, TX, 1993c.

Hatakeyarna, S., Pan, Y., and Furuta, K.: Identification of continuous systems via Markov parameters and time moments, *14th Triennial World Congress*, Beijing, 1999.

Haverkamp, B.R.J.: State space identification. PhD Thesis, Technical University of Delft, the Netherlands, 2001.

Haverkamp, B.R.J., Verhaegen, M., Chou, C.T., and Johansson, R.: Continuous-time subspace model identification method using Laguerre filters. *Proceedings of the 11th IFAC Symposium on System Identification*, Kitakyushu, Japan, 1997, pp. 1143–1148.

Haverkamp, B.R.J., Verhaegen, M., Chou, C.T., and Johansson, R.: Identification of continuous-time MIMO state space models from sampled data in the presence of process and measurement noise. *Proceedings of the 35th IEEE Conference on Decision Control* (*CDC*), Kobe, Japan, 1966, pp. 1539–1544.

Heuberger, P.S.C., van den Hof, P.M.J., and Wahlberg, B.: *Modelling and Identification with Rational Orthogonal Basis Functions*, Springer, London, 2005.

Hjalmarsson, H.: Aspects on incomplete modeling in system identification. PhD thesis, Department of Electrical Engineering, Linköping University, Linköping, Sweden, 1993.

Homssi, L., Titli, A., and Despujols, A.: Continuous-time process identification: Comparison of eight methods and practical aspects. *Proceedings of the 9th IFAC-Symposium on System Identification Parameter Estimation*, Budapest, Hungary, 1991, 2, pp. 1634–1642.

Hu, Y.Z., and Davison, E.J.: A PMGOP method of parameter identification for a class of continuous nonlinear systems, *Int. J. Contr.*, 1995, 62 (2), pp. 345–378.

Hunter, I.W., and Nielson, P.M.: Nonlinear and time-varying identification of living muscle cell mechanics. *Proceedings of the 11th IFAC World Congress*, Tallin, USSR, 1990, pp. 104–107.

Inoue, K., Kumamaru, K., Nakahashi, Y., Nakamura, N., and Uchida, M.: A quick identification method of continuous-time nonlinear systems and its application to power plant control. *Proceedings of the 10th IFAC Symposium on System Identification Parameter Estimation* (*SYSID94*), Copenhagen, Denmark, 1994, pp. 319–324.

Isermann, R., Lachmann, K.-H., and Matko, D.: *Adaptive Control Systems*, Prentice Hall, New York, 1992.

Jamshidi, M.: *Large-Scale Systems: Modelling and Control*. North-Holland, the Netherlands, 1983.

Jedner, U., and Unbehauen, H.: Identification of a class of nonlinear systems by parameter estimation of a linear multi-model. *Proceedings of the IMACS Symposium Modeling, Simulation, and Control Lumped Distributed Parameter System*, Villeneuve, France, 1986, pp. 297–299.

Jiang, Z.H., and Schaufelberger, W.: *Block Pulse Functions and Their Application in Control Systems*. Springer-Verlag, Berlin, Germany, 1992.

Johansson, R.: Identification of continuous-time models, *IEEE Trans. Signal Process.*, 1994, 42 (4), pp. 887–897.

Johansson, R., Verhaegen, M., Haverkamp, B.R.J., and Chou, C.T.: Correlation methods of continuous-time state-space model identification. *Proceedings of the European Control Conference* (*ECC*), Karlsruhe, Germany, 1999, paper F 1013-2.

Junge, T., and Unbehauen, H.: On-line identification of nonlinear systems using structurally adaptive radial basis functions. *Proceedings of the 16th American Control Conference* (*ACC97*), Albuquerque, NM, 1997, pp. 1037–1041.

Kabaila, P.V., and Goodwin, G.C.: On the estimation of the parameters of an optimal interpolator when the class of interpolators is restricted. *SIAM J Contr. Optim.*, 1980, 18 (2), pp. 121, 144.

Kalman, R.E.: *Aspects of Network and Systems Theory*. Holt, Reinhart and Winston, New York, 1970.

Kohr, R.H.: A method for determination of a differential equation model for simple nonlinear systems, *IEEE Trans. Electron. Comput.*, 1963, 12, pp. 394–400.

Kohr, R.H., and Haberock, L.L.: An experimental determination of differential equations to describe simple nonlinear systems. *Proceedings of the Joint Automatic Control Conference* (*JACC*), 1966, pp. 616–623.

Kollár, I.: *Frequency Domain System Identification Toolbox*. The Math-Works, Natick, MA, 1994.

Kortmann, P., and Unbehauen, H.: Structure identification of functional-type fuzzy models with application to modeling nonlinear dynamic plants, in Reusch, B. (Ed.): *Computational Intelligence*. LNCS 1226, Springer Verlag, Berlin, Germany, 1997.

Kosut, R.L.: Determining model uncertainty of identified models for robust control design. In *Preprints of the IFAC World Congress*, Sydney, Australia, 1993.

Kosut, R.L., Lau, M.K., and Boyd, S.P.: Set-membership identification of systems with parametric and nonparametric uncertainty. *IEEE Trans. Automat. Contr.*, 1992, 37 (7), pp. 929, 941.

Kreyszig, E.: *Introductory Functional Analysis with Applications*. John Wiley & Sons, New York, 1978.

Küper, P.: Identifikation kontinuierlicher dynamischer Systeme mit Hilfe von Markou-Parameters. Studienarbeit ESR 9138, Lehrstuhl für Elektrische Steuerung und Regelung, Ruhr Universität, Bochum, Germany, 1992.

Kurth, J.: *Identifikation nichtlinearer Systeme mit komprimierten Volterra-Reihen*. VDI-Verlag, Düsseldorf, Germany, 1995.

Larimore, W.: Canonical variate analysis identification, filtering and adaptive control. *Proceedings of the 29th IEEE-Conference on Decision Control* (*CDC*), Honolulu, Hawaii, 1990, pp. 596–604.

Li, L.M., and Billings, S.A.: Continuous-time linear and nonlinear system identification in the frequency domain. Research Report 711, Department Automatic Control and Systems Engineering, University of Sheffield, Sheffield, UK, 1998.

Lion, P.M.: Rapid identification of linear and nonlinear systems, *AIAA J.*, 1967, 5 (10), pp. 1835–1842.

Ljung, L.: Initialisation aspects for subspace and output-error identification methods. *Proceedings of the European Control Conference*, Cambridge, UK, CD-Rom Publication, 2003.

Ljung, L.: Some results on identifying linear systems using frequency domain data. *Proceedings of the 32nd IEEE Conference on Decision Control* (*CDC*), San Antonio, TX, 1993, pp. 3534–3538.

Ljung, L.: *System Identification: Theory for the User* (2nd ed.). Prentice Hall, Englewood Cliffs, NJ, 1999.

Ljung, L., and Söderström, T.: *Theory and Practice of Recursive Identification*. MIT Press, Cambridge, MA, 1983.

Ljungquist, D., and Balchen, J.: Recursive prediction error methods for online estimation in nonlinear state-space models. *Proceedings of the 32nd IEEE Conference on Decision and Control* (*CDC93*), San Antonio, CA, 1993, pp. 714–719.

Maciejowski, J.M.: *Multivariable Feedback Design*. Addison-Wesley Publishing Company, Reading, MA, 1989.

Mathew, A.V., and Fairman, F.W.: Identification in the presence of initial conditions, *IEEE Trans. Automat. Contr.*, 1972, 17, pp. 394–396.

Mehra, R.K.: On-line identification of linear dynamic systems with applications to Kalman filtering, *IEEE Trans. Automat. Contr.*, 1971, 16, pp. 12, 21.

Middleton, R.H., and Goodwin, G.C.: *Digital Control and Estimation: A Unified Approach*. Prentice Hall, Englewood Cliffs, NJ, 1990.

Milanese, M., and Vicino, A.: Estimation theory for dynamic systems with unknown but bounded uncertainty: An overview. *Proceedings of IFAC Symposium on Identification and System Parameter Estimation*, Budapest, Hungary, 1991.

Moonen, M., De Moor, B., and Vandevalle, J.: SVD-based subspace methods for multivariable continuous-time system identification, in Sinha, N., and Rao, G.P. (Eds.): *Identification of Continuous-time Systems*, Kluwer Academic Publishers, Dordrecht, the Netherlands, 1991.

Mukherjee, A.K., Saha, D.C., and Rao, G.P.: Identification of large scale distributed parameter systems—Some simplifications in the multidimensional Poisson moment functional approach, *Int. J. Sys. Sci.*, 1983, 14 (7), pp. 777–788.

Mukhopadhyay, S.: Continuous-time models and approaches for estimation and control of linear systems. PhD thesis, Department of Electrical Engineering, Indian Institute of Technology, Kharagpur, India, 1990.

Mukhopadhyay, S., Patra, A., and Rao, G.P.: Irreducible model estimation for MIMO systems, *Int. J. Contr.*, 1991, 53 (1), pp. 223, 253.

Mukhopadhyay, S., Patra, A., and Rao, G.P.: A new class of discrete-time models for continuous-time systems, *Int. J. Contr.*, 1992, 55 (5), pp. 1161–1187.

Mukhopadhyay, S., and Rao, G.P.: Integral-equation approach to joint state and parameter estimation in continuous-time MIMO systems. *Proceedings of IEE*, Part D, 1991, 138 (2), pp. 93, 102.

Neural Network Identification of Nonlinear Systems Toolbox for Use with MATLAB. Scientific Computers GmbH, Aachen, Germany, Franzstr. 107.

Nagpal, K.M., and Khargonekar, P.P.: Filtering and smoothing in H setting. *IEEE Trans. Automat. Contr.,* 1991, 36 (2), pp. 152, 166.

Niederlinski, A., and Hajdasinski, A.: Multivariable system identification—A survey. *Proceedings of IFAC Symposium on Identification and System Parameter Estimation,* pp. 43, 76, 1979.

Norton, J.P.: Identification and application of bounded parameter models, *Automatica,* 1987a, 23 (4), pp. 497, 507.

Norton, J.P.: Identification of parameter bounds for ARMAX models from records with bounded noise, *Int. J. Contr.,* 1987b, 45 (2), pp. 375, 390.

Ohsumi, A., Kameyama, K., and Yamaguchi, K.: Subspace-based identification for continuous-time stochastic systems via distribution-theoretic approach. *Proceedings of the 14th IFAC World Congress,* Beijing, 1999, pp. 223–228.

Padilla, A., Garnier, H., and Gilson, M.: Version 7.0 of the CONTSID toolbox. *17th IFAC Symposium on System Identification, SYSID 2015,* October 2015, Beijing, 2015.

Patra, A., and Rao, G.P.: *General Hybrid Orthogonal Functions and their Applications in Systems and Control,* LNCIS-213, Springer, Berlin, Germany, 1996.

Patra, A., and Unbehauen, H.: Identification of a class of nonlinear continuous-time systems using Hartley modulating functions, *Int. J. Contr.,* 1995, 62 (6), pp. 1431–1451.

Pazdera, J.S., and Pottinger, H.J.: Linear system identification via Liapunov design techniques. *Proceedings of the Joint Automatic Control Conference,* Boulder, CO, 1969, pp. 795–801.

Pearson, A.E.: Explicit parameter identification for a class of nonlinear input/output differential operator models. *Proceedings of the 31st IEEE Conference on Decision and Control* (CDC), Tucson, AZ, 1992, pp. 3656–3660.

Pearson, A.E.: Least squares parameter identification of nonlinear differential input-output models. *Proceedings of the 27th IEEE Conference on Decision and Control* (CDC), Austin, TX, 1988, pp. 1831–1835.

Pearson, A.E., and Lee, F.C.: On the identification of polynomial input-output differential systems, *IEEE Transactions on Automatic Control,* 1985, 30 (8), pp. 778–782.

Pearson, A.E., and Lee, F.C.: Parameter identification of linear differential systems via Fourier based modulating functions, *Control Theory and Advanced Technology (C-TAT),* 1985, 1, pp. 239–266.

Pearson, A.E., Shen, Y., and Klein, V.: Application of Fourier modulating functions to parameter estimation of a multivariable linear differential system. *Proceedings of the 10th IFAC Symposium System Identification* (*SYSID94*), Copenhagen, Denmark, 1994, pp. 49–54.

Pearson, A.E., Shen, Y., and Pan, J.Q.: Discrete frequency formats for linear differential system identification. *Proceedings of the 12th IFAC World Congress,* Sydney, Australia, 1993, VII, pp. 143–148.

Pintelon, R., Guillaume, P., Rolain, Y., Schoukens, J., and Van Hamme, H.: Parametric identification of transfer functions in the frequency domain—A survey, *IEEE Transactions on Automatic Control,* 1994, 39 (11), pp. 2245–2260.

Pottmann, M., Unbehauen, H., and Seborg, D.: Application of a general multi-model approach for identification of highly nonlinear processes—A case study, *Int. J. Contr.,* 1993, 57, pp. 97–120.

Rao, G.P.: *Piecewise Constant Orthogonal Functions and Their Application to Systems and Control*, LNCIS Vol. 55. Springer Verlag, Berlin, Germany, 1983.
Rao, G.P., Diekmann, K., and Unbehauen, H.: Parameter estimation in large-scale interconnected systems. *Proceedings of the 9th IFAC World Congress*, Budapest, Hungary, 1984, X, pp. 118–122.
Rao, G.P., Diekmann, K., and Unbehauen, H.: Parameter estimation in large-scale interconnected systems. *Proceedings of IFAC Symposium on Identification and System Parameter Estimation*, Budapest, Hungary, 1984, pp. 729, 733.
Rao, G.P., and Garnier, H.: Identification of continuous-time systems: Direct or indirect? *Proceedings of the 15th International Conference on Systems Science*, Wroclaw, Poland, 2004, I, pp. 66–86.
Rao, G.P., and Garnier, H.: Numerical illustrations of the relevance of direct continuous-time model identification. *15th Triennial IFAC World Congress*, Barcelona, Spain, 2002, CD-Rom publication (paper 2456).
Rao, G.P., Saha, D.C., Rao, T.M., Bhaya, A., and Aghoramurthy, K.: A microprocessor-based system for on-line parameter identification in continuous dynamical systems, *IEEE Trans. Ind. Electron.*, 1982, 29 (3), pp. 197–201.
Rao, G.P., and Sivakumar, L.: Identification of deterministic time lag systems, *IEEE Trans. Automat. Contr.*, 1976, 21, pp. 527–529.
Rao, G.P., and Sivakumar, L.: Parameter identification in lumped linear continuous systems in a noisy environment via Kalman filtered Poisson moment functionals, *Int. J. Contr.*, 1982, 35 (3), pp. 509–519.
Rao, G.P., and Sivakumar, L.: System identification via Walsh functions, *Proceedings of IEE*, 1975, 122 (10), pp. 1160–1161.
Rao, G.P., and Srinivasan, T.: Remarks on "Authors reply" to "Comments on the design of piecewise constant gains for optimal control," *IEEE Trans. Automat. Contr.*, 1978, 23 (4), pp. 762–763.
Rao, G.P., and Subrahmanyam, A.V.B.: Models in generalized MA form for identification of continuous-time systems, in Katayama, T., and Sugimoto, S. (Eds.): *Statistical Methods in Control and Signal Processing*, Marcel Dekker, New York, 1996.
Rao, G.P., and Tzafestas, S.G.: A decade of piecewise constant orthogonal functions in systems and control, *Math. Comput. Simul.*, 1985, 27 (5), pp. 389–407.
Rao, G.P., and Unbehauen, H. (Guest Editors): *C-TAT, Special Issue on Identification and Adaptive Control—Continuous Time Approaches*, 1993.
Rao, G.P., and Unbehauen, H.: Identification of continuous-time systems. *Proceedings of IEE Control Theory and Applications*, 2006, 153 (2), pp. 185–220.
Rao, G.P., Unbehauen, H., Mukhopadhyay, S., and Patra, A.: From calculus to algebra in models of continuous-time systems. *Proceedings of the 9th IFAC/IFORS Symposium on Identification and System Parameter Estimation*, Budapest, Hungary, 1991, 2, pp. 1340–1346.
Read, K.N., and Ray, W.H.: Application of nonlinear dynamic analysis in the identification and control of nonlinear systems—I. Simple dynamics, *J. Process Contr.*, 1998, 8 (1), pp. 1–15.
Rolain, Y., Pintelon, R., Xu, K.Q., and Vold, H.: On the use of orthogonal polynomials in high order frequency domain system identification and its application to model parameter estimation. *Proceedings of the 33rd IEEE Conference on Decision and Control* (CDC), Lake Buena Vista, FL, 1994, pp. 3365–3373.

Rucker, R.A.: Real-time system identification in the presence of noise. Pepr. IEEE Western Electronics Convention, 1963, paper 2.9.

Rufer, D.F.: Implementation and properties of a method for identification of nonlinear continuous-time models. Fachgruppe Automatik, ETH Zürich, Report 77-01, 1977.

Sagara, S., Yang, Z.J., and Wada, K.: Identification of continuous systems from noisy sampled input-output data. *Proceedings of the 9th IFAC/IFORS Symposium on Identification and System Parameter Estimation*, Budapest, Hungary, 1991, pp. 603–608.

Sagara, S., Yang, Z.J., and Wada, K.: Identification of continuous systems using digital low-pass filtering, *Int. J. Sys. Sci.*, 1991, 22 (7), pp. 1159–1179.

Sagara, S., Yang, Z.J., and Wada, K.: Recursive identification algorithms for continuous systems using an adaptive procedure, *Int. J. Contr.*, 1991, 53 (2), pp. 391–409.

Sagara, S., and Zhao, Z.Y.: Filtering approaches to identification of continuous systems. *Proceedings of the 9th IFAC/IFORS Symposium Identification and System Parameter Estimation*, Budapest, Hungary, 1991, pp. 305–310.

Sagara, S., and Zhao, Z.Y.: Numerical integration approach to on-line identification of continuous-time systems in the presence of measurement noise. *Proceedings of the 8th IFAC/IFORS Symposium Identification and System Parameter Estimation,* Beijing, 1988, pp. 441–446.

Sagara, S., and Zhao, Z.Y.: Recursive identification of transfer function matrix in continuous systems via linear integral filters, *Int. J. Contr.*, 1989, 50 (2), 457, 477, 1989.

Saha, D.C., and Rao, G.P.: A general algorithm for parameter identification in lumped continuous systems: The Poisson moment functional (PMF) approach, *IEEE Trans. Automat. Contr.*, 1982, 27 (1), pp. 223–225.

Saha, D.C., and Rao, G.P.: *Identification of Continuous Systems: The Poisson Moment Functional Approach.* Lecture Notes in Control and Information Sciences-56, Springer-Verlag, Berlin, Germany, 1983.

Saha, D.C., and Rao, G.P.: Identification of lumped linear systems in the presence of initial conditions via Poisson moment functionals, *Int. J. Contr.*, 1980a, 31 (4), pp. 637–644.

Saha, D.C., and Rao, G.P.: Identification of lumped linear systems in the presence of small unknown time delays—The Poisson moment functional approach, *Int. J. Contr.*, 1981, 33 (5), pp. 945–951.

Saha, D.C., and Rao, G.P.: Identification of lumped linear time varying parameter systems via Poisson moment functionals, *Int. J. Contr.*, 1980b, 32 (4), pp. 709–721.

Saha, D.C., and Rao, G.P.: Transfer function matrix identification in MIMO systems via Poisson moment functionals, *Int. J. Contr.*, 1982, 35 (4), pp. 727–738.

Saha, D.C., Rao, B.B.P., and Rao, G.P.: Structure and parameter identification in linear continuous lumped systems—The Poisson moment functional approach, *Int. J. Contr.*, 1982b, 36 (3), pp. 477–491.

Schaufelberger, W.: Modelladaptive systeme. Dr.-Ing. Dissertation, ETH, Zürich, Switzerland, 1969.

Schoukens, J., and Pintelon, R.: *Identification of Linear Systems: A Practical Guideline to Accurate Modeling.* Pergamon Press, London, UK, 1991.

Schoukens, J., Pintelon, R., and Van Hamme, H.: Identification of linear dynamic systems using piecewise constant excitations: Use, misuse and alternatives, *Automatica*, 1994, 30 (7), pp. 1153–1169.

Schoukens, J., Pintelon, R., Vandersteen, G., and Guillaume, P.: Frequency-domain system identification using non-parametric noise models estimated from a small number of data sets, *Automatica*, 1997, 33 (6), pp. 1073–1086.

Shaked, U., and Theodor, Y.: $\mathcal{H}_\infty$-optimal estimation: A tutorial. *Proceedings of IEEE Conference on Decision and Control*, Tucson, AZ, 1992.

Shinbrot, M.: On the analysis of linear and nonlinear systems, *Trans. ASME*, 1957, 79, pp. 547–552.

Sinha, N.K., Mahalanabis, A.K., and El-Sherief, H.: A nonparametric approach to the identification of linear multivariable systems, *Int. J. Syst. Sci.*, 1978, 8, pp. 425, 430.

Sinha, N.K., and Rao, G.P., (Eds.): *Identification of Continuous Time Systems: Methodology and Computer Implementation*. Kluwer Academic Publishers, Dordrecht, the Netherlands, 1991.

Skantze, F.P., Pearson, A.E., and Klein, V.: Parameter identification for unsteady aerodynamic systems by the modulating function technique. *Proceedings of 10th IFAC Symposium on Syst. Identification* (*SYSID94*), Copenhagen, Denmark, 1994, pp. 571–576.

Söderström, T., Fan, H., Carlsson, B., and Bigi, S.: Least squares parameter estimation of continuous-time ARX models from discrete-time data, *IEEE Trans. Automat. Contr.*, 1997, 42 (5), pp. 659–673.

Söderström, T., and Stoica, P.: *System Identification*. Prentice-Hall, New York, 1989.

Subrahmanyam, A.V.B., and Rao, G.P.: Identification of continuous-time SISO systems via Markov parameter estimation. *Proceedings of IEE*, Part D, 1993, 140(1), pp. 1, 10.

Subrahmanyam, A.V.B., Saha, D.C., and Rao, G.P.: Identification of continuous-time MIMO systems via time moments, *Contr. Theory Adv. Technol.*, 1995, 10 (4), pp. 1359–1378.

Subrahmanyam, A.V.B., Saha, D.C., and Rao, G.P.: Irreducible continuous model identification via Markov parameter estimation, *Automatica*, 1996, 32 (2), pp. 249–253.

Sun, W., Nagpal, K.M., and Khargonekar, P.P.: $\mathcal{H}_\infty$ control and filtering for sampled data systems, *IEEE Trans. Automat. Contr.*, 1993, 38 (8), pp. 1162, 1175.

Tether, A.: Construction of minimum state-variable models from finite input-output data, *IEEE Trans. Automat. Contr.*, 1970, 15 (4), pp. 427–436.

Tsang, K.M., and Billings, S.A.: Identification of continuous time nonlinear systems using delayed state variable filters, *Int. J. Contr.*, 1994, 60 (2), pp. 159–180.

Unbehauen, H.: Adaptive model approaches, in Sinha, N., and Rao, G.P. (Eds.): *Identification of Continuous-time Systems*, Kluwer Academic Publishers, Dordrecht, the Netherlands, 1991.

Unbehauen, H.: Some new trends in identification and modelling of nonlinear dynamical systems, *J. Appl. Math. Comput.*, 1996, 78, pp. 279–297.

Unbehauen, H., and Rao, G.P.: A review of identification in continuous systems, *IFAC Annu. Rev. Contr.*, 1998, 22, pp. 145–171.

Unbehauen, H., and Rao, G.P.: Continuous-time approaches to system identification—A survey, *Automatica*, 1990, 26 (1), pp. 23–35.

Unbehauen, H., and Rao, G.P.: *Identification of Continuous Systems*. North Holland, Amsterdam, the Netherlands, 1987.

Unbehauen, H., and Rao, G.P.: Identification of continuous systems—A survey, *Syst. Anal. Model. Simul.*, 1998, 33, pp. 99–155.

Unbehauen, H., and Rao, G.P.: Identification of continuous-time systems—A tutorial. *Proceedings of the 11th IFAC Symposium on Identification (SYSID97)*, Fukuoka, Japan, July 1997, pp. 1023–1049.

Ungarala, S., and Co, T.B.: Time-varying system identification using modulating functions and spline models with applications to bio-processes, *Comput. Chem. Eng.*, 2000, 24, pp. 2739–2753.

van den Boom, T.J.J., Klompstra, M., and Damen, A.A.H.: System identification for $\mathcal{H}\infty$-robust control design. *Proceedings of IFAC Symposium on Identification and System Parameter Estimation*, Budapest, Hungary, 1991.

Van Overschee, P., and De Moor, B.: N4SID: Subspace algorithms for the identification of combined deterministic stochastic systems, *Automatica*, 1994, 30 (1), pp. 75–93.

Van Overschee, P., and De Moor, B.: *Subspace Identification for Linear Systems*, Kluwer Academic Publishers, Dordrecht, the Netherlands, 1996.

Verhaegen, M.: A novel non-iterative MIMO state space model identification technique. *Proceedings of the 9th IFAC/IFORS Symposium on Identification and System Parameter Estimation*, Budapest, Hungary, 1991, pp. 1453–1458.

Verhaegen, M.: Identification of the deterministic part of MIMO state space models given in innovations form from input-output data, *Automatica*, 1994, 30 (1), pp. 61–74.

Verhaegen, M.: Subspace model identification, Pt. 3: Analysis of the ordinary output-error state space model identification algorithm, *Int. J. Contr.*, 1993, 58, pp. 555–586.

Wahlberg, B.: Identification of resonant systems using Kautz filters. *Proceedings of IEEE Conference on Decision and Control*, Brighton, UK, 1991.

Wahlberg, B.: Laguerre and Kautz models. *Proceedings of the 10th IFAC Symposium on Identification and System Parameter Estimation (SYSID94)*, Copenhagen, Denmark, 1994, 3, pp. 1–12.

Wahlberg, B., and Ljung, L.: Design variables for bias distribution in transfer function estimation, *IEEE Trans. Automat. Contr.*, 1986, 31 (2), pp. 134–144.

Wahlberg, B., and Ljung, L.: Hard frequency-domain model error bounds from least-squares like identification techniques. *IEEE Trans. Automat. Contr.*, 1992, 37 (7), pp. 900, 912.

Wang, S.: Block pulse operator method for parameter identification problems in non-linear continuous systems, *Int. J. Syst. Sci.*, 1991, 22 (12), pp. 2441–2455.

Wang, S.-Y.: On the orthogonal functions approximation method of non-linear system identification and its application in biochemical processes, *Contr. Theory Adv. Technol.*, 1993, 9 (1), pp. 39–52.

Whitfield, A.: Transfer function synthesis using frequency response data, *Int. J. Contr.*, 1986, 43 (5), pp. 1413–1425.

Widrow, B., and Stearns, S.D.: *Adaptive Signal Processing*. Prentice Hall, Englewood Cliffs, NJ, 1985.

Wiener, N.: *Nonlinear Problems in Random Theory*. MIT Press, Cambridge, MA, 1958.

Wolovich, W.A.: *Linear Multivariable Systems*. Springer Verlag, New York, 1974.

Yang, Z.Y., Sagara, S., and Wada, K.: Identification of continuous-time systems from sampled input-output data using bias eliminating techniques, *Contr. Theor. Adv. Technol.*, 1993, 9 (1), pp. 53, 75.

Yang, Z.J., Sagara, S., and Wada, K.: Parameter identification based on the Steiglitz-McBride method from noisy input-output data, *Trans. Soc. Instrum. Contr. Eng.*, 1992, 28 (12), pp. 1492–1494.

Yang, Z.Y., Sagara, S., and Wada, K.: Subspace model identification for continuous-time systems. *Proceedings of the 11th IFAC Symposium on System Identification and Parameter Estimation*, Kitakyushu, Japan, 1997, pp. 1671–1676.

Young, P.C.: An instrumental variable method for real-time identification of a noisy process, *Automatica*, 1970, 17, pp. 23, 39.

Young, P.C.: Applying parameter estimation to dynamic systems: Part 2, *Contr. Eng.*, 1969, 16 (11), pp. 118–124.

Young, P.C.: Comments on "On the estimation of continuous-time transfer functions," *Int. J. Contr.*, 2002, 75, pp. 693–697.

Young, P.C.: In flight dynamic checkout—A discussion, *IEEE Trans. Aerospace Electron. Syst.*, 1964, 2, pp. 1106–1111.

Young, P.C.: Parameter estimation of continuous-time models—A survey, *Automatica*, 1981, 17 (1), pp. 23–39.

Young, P.C.: Process parameter estimation and self adaptive control, in Hammond, P.H. (Ed.): *Theory of Self Adaptive Control Systems*, Plenum Press, New York, 1965, pp. 118–140.

Young, P.C.: The determination of the parameters of a dynamic process, *Radio Electron. Eng. J. IERE*, 1965, 29, pp. 345–361.

Young, P.C., Chotai, A., and Tych, W.: Identification, estimation and control of continuous-time systems described by delta operator models, in Sinha, N.K., and Rao, G.P. (Eds.): *Identification of Continuous-time Systems*, Kluwer, Dordrecht, the Netherlands, 1991, pp. 363–418.

Young, P., and Garnier, H.: The advantages of directly identifying continuous-time transfer function models in practical applications, *Int. J. Contr.*, 87 (7), 2014.

Young, P.C., Garnier, H., and Javis, A.: Identification of continuous time linear and nonlinear models: A tutorial with environmental applications. *Proceedings of the 13th IFAC Symposium on System Identification*, Rotterdam, the Netherlands, 2003, pp. 618–628.

Young, P.C., and Jakeman, A.L.: Refined instrumental variable methods of recursive time-series analysis: Part III, Extensions, *Int. J. Contr.*, 1980, 31, pp. 741–764.

Young, P.C., McKenna, P., and Bruun, J.: Identification of nonlinear stochastic systems by state dependent parameter estimation, *Int. J. Contr.*, 2001, 74, pp. 1837–1857.

Zadeh, L.A.: From circuit theory to system theory, *Proc. Inst. Radio Eng.*, 1962, 50, pp. 856–865.

Zames, G.: Feedback and optimal sensitivity: Model reference transformations, multiplicative seminorms, and approximate in verses, *IEEE Trans. Automat. Contr.*, 1981, 26, pp. 301, 320.

Zhao, Z.Y., Sagara, S., and Kumamaru, K.: On-line identification of time delay and system parameters of continuous systems based on discrete-time measurements. *Proceedings of the 9th IFAC/IFORS Symposium on Identification and System Parameter Estimation*, Budapest, Hungary, 1991a, pp. 721–726.

Zhao, Z.Y., Sagara, S., and Tomizuka, M.: A new bias compensating least-squares method for continuous system identification in the presence of coloured noise, *Int. J. Contr.*, 1992, 56 (6), pp. 1441, 1452.

Zhao, Z.Y., Sagara, S., and Wada, K.: Bias-compensating least squares method for identification of continuous-time systems from sampled data, *Int. J. Contr.*, 1991b, 53 (2), pp. 445, 461.

Subject Index

Author Index

I

J

K

L

M

N